P9-DFN-541

With love,

to Jot

It Started with Copernicus

How Turning the World Inside Out Led to the Scientific Revolution

Howard Margolis

McGraw-Hill
New York Chicago San Francisco
Lisbon London Madrid Mexico City Milan
New Delhi San Juan Seoul Singapore
Sydney Toronto

Library of Congress Cataloging-in-Publication Data

Margolis, Howard.
It started with Copernicus : how turning the world inside out led to the scientific revolution / Howard Margolis.
p. cm.
Includes bibliographical references and index.
ISBN 0-07-138507-X
1. Science, Medieval. 2. Science, Renaissance. I. Title.

Q124.97 .M37 2002
509.4'09'032—dc21 2001055829

McGraw-Hill

A Division of The ***McGraw-Hill*** *Companies*

1 2 3 4 5 6 7 8 9 0 AGM/AGM 0 8 7 6 5 4 3 2

ISBN 0-07-138507-X

Printed and bound by Quebecor / Martinsburg.

This book is printed on recycled, acid-free paper containing a minimum of 50% recycled de-inked fiber.

It Started with Copernicus

Other Books by Howard Margolis

Dealing with Risk: Why Expert and Lay Intuition Conflict, and What Might Be Done about It

Paradigms and Barriers: How Habits of Mind Govern Scientific Belief

Patterns, Thinking, and Cognition: A Theory of Judgment

Selfishness, Altruism, and Rationality: A Theory of Social Choice

Contents

Acknowledgments

I have asked too many people too many questions over the course of this project to hope to list them all. But I need to mention at least Susan Abrams, Wilbur Applebaum, Norman Bradburn, Michael Fisher, Daniel Garber, Owen Gingerich, Leslie Kurke, Anna Lussardi, Bob Richards, Eric Schliesser, Stephen Stigler, Bruce Stephenson, Noel Swerdlow, John Tryneski, and Geert Vanpaemel.

The argument is often developed with the help of illustrations from influential books of the sixteenth and early seventeenth centuries. My thanks to the University of Chicago Library, whose rare book collection has proved to be an important asset throughout the project.

Amy Bianco, who acquired this project for McGraw-Hill, provided much advice for which I am grateful, and for which readers would be grateful had they experienced the alternative.

Finally I owe my largest debt to Thomas Kuhn. Neither this book nor my earlier work on cognition could have been written other than under the influence of Kuhn's celebrated *Structure of Scientific Revolutions*. And I may not have enjoyed the comfortable situation from which I was able to write it without his support.

I thank Jehane Kuhn for permission to use excerpts from Kuhn's letters to me of October 23 and November 8, 1988.

* * *

Supplementary information related to this book will be posted at: http://www.harrisschool.uchicago.edu/wp/02-02.html.

It Started with Copernicus

Introduction

If you read what has been written over the past half century about Copernicus, or what has been written about the Scientific Revolution, you will have some doubt that anything of deep importance started with Copernicus, least of all something that can be properly labeled "the Scientific Revolution."[1] But if you look closely at what was going on in science around the year 1600, you will have no trouble seeing the appropriateness of a story of the Old West about a cowboy wandering over the plateau of northern Arizona. Innocently, he rides right up to the rim of the Grand Canyon. The cowboy sits for a long time, contemplating the vast gorge. Eventually he mutters, "Something happened here." And you need not look far to notice a powerful hint that what happened in physics c. 1600 was somehow linked to Copernicus.

Until the end of the sixteenth century, it would make some sense to claim that things changed in physics rather as things changed in fashion. A knowledgeable historian might be able to assign a sample of early scientific work to the right place and period. But the physics taught to Galileo—or even the physics Galileo himself was teaching into the 1590s—was only doubtfully superior to what students could have learned from Archimedes 1800 years earlier. And then, suddenly, over a period of a few years around 1600, something happened.

Physics moved decisively ahead. Poetry after 1600 was not unambiguously better than all earlier poetry. Shakespeare does not wholly dominate Dante or Homer. But physics, and science in general, now was unmistakably better.

The pace of discovery changed so dramatically that we could say that there was somehow a discovery of discovery. A reader who is skeptical of this claim might trying filling in the right side of Table I-1. For a reason that will soon become very apparent, the table leaves off the Copernican claim that the Earth revolves around the Sun, which was published in 1543. So the question to ponder is this: Leaving aside Copernicus, is there any discovery *in science* over the fourteen centuries prior to about 1600 that is as significant as any of the items listed for c. 1600?

This category of discovery "in science" is not intended to cover everything: Technological discovery (printing, gunpowder, clocks, and much more) is not there, nor is mathematical discovery (notably algebra). We are concerned here only with discoveries in pure science. But modern life depends absolutely on what first emerged as discoveries in pure science. We can see what is happening on the far side of the Earth, visit other planets, live twice as long, fly through the air, and in general take for granted a world of things that not so long ago would have been considered possible only by miracle or magic. And what we want to understand is why the flow of discovery in pure science on which all this depends began so abruptly about 1600.

The "previous fourteen centuries" on the blank right side of the table makes the time frame for prior to 1600 extend back to right after Galen and Ptolemy, c. 150 A.D. And the list on the left does not cover all remarkable discoveries c. 1600, but only discoveries (in natural science) that proved sufficiently memorable that any reasonably well-informed person might immediately know what they are.[2] Of course there is room for interpretation here. But relevant candidates for the "previous 1400 years" column would have to be such that in a rank ordering of importance, they could plausibly come somewhere above the very bottom of the "c. 1600" list.[3]

Table I-1 Notable Scientific Discoveries

Made c. 1600	Made in the Previous Fourteen Centuries
• Distinction between electricity and magnetism	
• Law of free fall	
• Galilean inertia	
• Earth is a magnet	
• Theory of lenses	
• Laws of planetary motion	
• Various discoveries with the telescope (the Moon is Earthlike, with mountains; the Sun exhibits spots and rotates; Jupiter has moons; etc.)	
• Laws of hydrostatic pressure	
• Synchronicity of the pendulum	

I have been presenting this table to skeptical historians for some time without eliciting any strong counterexample (I later discuss the best candidates I have been able to find). But on the most generous account, what might be produced could hardly provide a rebuttal of what the table shouts out very loudly: something happened at just about 1600. The slow pace of discovery in classical science faded to close to zero after the time of Ptolemy. There were brilliant investigators over those many centuries. But to a remarkable extent, even brilliant investigators produced very little in the way of brilliant discoveries for nearly 1500 years. Then, at around the turn of the seventeenth century, with a decided abruptness, the dam burst. Something happened that started what we have come to call the Scientific Revolution.

What Happened?

But what was it that happened? Historians recently have been full of doubt that there is in fact anything revolutionary to be found. A recent book aimed at history of science courses hit a fashionable note when it opened with the remark: "There was no such thing as the Scientific Revolution, and this is a book about it."

But that nothing has been noticed can't refute what Table I-1 bluntly shows. Whether we have identified what happened or not, plainly something did happen. On the other hand, that recent writing so often doubts there is anything to be noticed implies a certain invisibility of whatever in fact was new. Apparently whatever was new then is now so commonplace that it is hard to imagine its absence.

The most obvious proposal does not work. The Scientific Revolution can't be explained as just a matter of scientists coming to understand that ideas about nature had to be tested by experiments. Even though scientists later in the seventeenth century self-consciously regarded themselves as practitioners of a "new philosophy" that was synonymous with "experimental philosophy, "a turn to experiment without something more fails to account for what happened because so much evidence has been found of experiments prior to 1600.[4] Experiments were not as common before 1600 as they would soon become, but they clearly played a role all the way back to classical times.[5] Nor could an emphasis on mathematics account for the change. Until we reach Huygens and Newton much later in the seventeenth century, mathematical techniques in science never went conspicuously beyond what can be seen in Archimedes and Ptolemy.

So if there was something new to be identified—as Table I-1 shouts out loud that there surely must be—it wasn't experimenting or even the elaborateness of experimenting, nor was it the use of more sophisticated mathematics. And just "more" of what had always been on hand would not explain what happened unless something could be pointed to that had prompted it. The sharp step that is visible at about 1600 almost by definition requires something really new. We need a bit of discovery to find out what that was.

For a while historians tried spreading the Scientific Revolution over two centuries or longer. The Scientific Revolution became a part of the broad resurgence of European culture and economy that began in the late Middle Ages. That is most certainly part of the story, but it is not enough to explain the abruptness of the burst of discovery c. 1600. Table I-1 tells us that to neglect the sharp turn at the beginning of the seventeenth century is to miss the puzzle, not to solve it. And seeing the Scientific Revolution simply as a continuation of the broad cultural developments of the Renaissance does not begin to explain another decidedly odd feature: Every one of the discoveries in Table I-1 came from just four people, and the four shared a remarkable characteristic amongst themselves but with no one else

Stevin, Gilbert, Kepler, and Galileo

The discoveries in Table I-1 were produced by men who lived in four different countries, followed four different religions, and had four different primary occupations. They all eventually held positions at the courts of their respective countries, but different positions.[6] Stevin, Gilbert, Kepler, and Galileo, however, shared one very rare characteristic: At a time when few people were Copernican, each was not only a Copernican but a sufficiently enthusiastic Copernican to write a technical book expounding a Copernican view of the heavens. Stevin, Kepler, and Galileo were aggressively Copernican. Gilbert was decidedly careful. He goes out of his way to be sure a reader notices that he is not *saying* that he is Copernican. But then he is not shy at all about *showing* he is Copernican.[7]

So the burst of discovery around 1600 came from three men who were outspokenly Copernican plus a fourth who was cautious but still repeatedly comments on details of the world system in a way that could make sense only for a Copernican. So perhaps what enabled these four to produce a burst of discovery was something that a person might learn from Copernicus about how to make discoveries. And indeed, the key Copernican argument that we will have occasion to review in Chapter 3 turns out to be something that was logically at hand for every astronomer since Ptolemy. Somehow Copernicus had come to see something astonishing that astronomers before him had missed for fourteen centuries.

Sharpening the Puzzle

But many of the discoveries c. 1600 (Kepler's work on optics, Stevin and Galileo's "hydrostatic paradox") involve nothing Copernican. The most famous discovery to immediately follow those in the table was the discovery of the circulation of the blood, which came from Gilbert's eventual successor as royal physician, William Harvey. Not only does this have nothing to do with Copernicanism, but Harvey himself does not appear to have been Copernican. And more generally, what we see is not an isolated burst of discovery, but a radical and permanent change in the pace of discovery. Apparently it was not the Copernican idea per se but some change in method of discovery that was crucial. Somehow the four Copernicans active around 1600 collectively provided a model for future discovery.

Yet Copernicus himself was certainly no revolutionary in method. This is so obvious that in recent historical writing he is commonly treated as a conservative who held back real progress in astronomy. "The timid canon of Frauenberg" was Koestler's label. *De Revolutionibus* was published as Copernicus lay on his deathbed, in 1543. How could this not only inspire a revolution in method (though Copernicus himself was certainly no revolutionary in that respect) but exert some large methodological influence after a lapse of more than half a century?[8] On the other hand, how likely could it be that the most astounding scientific claim that had ever been made would have nothing to do with a burst of further surprising discoveries from just the people who were convinced that the Copernican claim was right?

I try to provide a single coherent account that fits all these pieces together. Something happened. But what? The tale unfolds to yield a resolution akin to Poe's "Purloined Letter": What happened is something that we all knew about already, but that today we take so much for granted that it is invisible.

Notes

1. Here is a sample:

   "Had it not been for Tycho Brahe and Kepler, the Copernican system would [only] have contributed to the perpetuation of the Ptolemaic system in a slightly more complicated form" (Otto Neugebauer, 1958, p. 505).

   "The timid canon of Frauenberg" (Copernicus as seen by Arthur Koestler, 1965).

   "The idea that there was a Copernican revolution in science . . . is an invention of later historians" (I. B. Cohen, in his Pfizer Prize book on scientific revolutions, 1985, p. 106).

   "Until the advent of the telescope, at the very earliest, the available evidence did not render belief in the mobility of the Earth even plausible, let alone convincing " (Victor Thoren, in his biography of Tycho, 1990).

   "There was no such thing as the Scientific Revolution" (Steven Shapin opening his book on the Scientific Revolution, 1996).

   "The Scientific Revolution is longer a historical event" (From the article "Scientific Revolution" in the Cambridge *Reader's Guide to the History of Science*).
2. So what was really a fundamental discovery by Kepler on the formation of pinhole images (discussed in Chapter 5) has been left off the list.
3. By far the most common suggestion I have encountered points to the Islamic work by Alhazen and others on intromission theory around the twelfth century. But Aristotle supported the intromission theory, so something more specific is needed. For reasons that I spell out in Chapter 4, even this salient case is doubtful.
4. See Duhem (1905, 1991).
5. See Cohen and Drabkin's *Source Book in Greek Science*, especially the sections on mechanics and optics. This is a modest gleaning considering

that the book covers a time span of about six centuries. On the other hand, it makes it plain that careful experiments were done long before 1600.

6. Stevin, who was primarily what we would call an engineer, served as tutor and then adviser to the prince. Gilbert was physician to Queen Elizabeth and then to James I. Kepler eventually held the title of mathematician to the Holy Roman Emperor. His contacts with the court consisted mostly of pleading for his salary. Galileo was a professor of mathematics until his discoveries with the telescope won him appointment as chief mathematician and philosopher to the duke of Florence.
7. Gilbert is usually treated as no more than leaning toward full Copernicanism. In both *De Magnete* and *De Mundo*, he repeatedly reminds the reader that beyond the daily rotation, there is also the orbital motion of the Earth to be considered. But each time he raises the question only to remind the reader that he is not answering it. Yet numerous other remarks in both books make little sense by themselves and no sense at all in the aggregate unless his unstated answer is Copernican. He casually refers (in *De Magnete*) to "the Earth and the other planets," and he remarks, "I pass over the reasons of the Earth's remaining motions, for at present the only question is concerning its diurnal rotation" (trans. Gilbert,1958, p. 220). Here and in numerous in other passages, he says or implies that the Earth must be a planet. He is emphatic in both *De Magnete* and *De Mundo* on treating the Earth and the Moon as very similar bodies. He even refers to the Earth's annual motion in "a great orbit" (p. 232). In the end (see the further evidence in Chapter 5) there is no more reason to doubt that Gilbert in fact was Copernican than to doubt that Darwin saw man as a product of evolution when he wrote the *Origin of Species*, or indeed that Galileo was Copernican though throughout his *Dialogue* (like Darwin on human evolution in the *Origin*) he avoided saying so. What might account for Gilbert's tactics? One point to note is that Copernicus's own book was printed with an unsigned preface telling its readers that they should not take the claims about the Earth's motion literally. We now know the preface was added by a cautious editor who was seeing the manuscript through the press, but Gilbert had no way of knowing that. So there was a distinguished precedent for Gilbert's *De Magnete* coming out with a comparable protective preface. This defends Gilbert's aggressive argument for the daily rotation of the Earth as consistent with the Bible, but in a way that suggests that the orbital motion would not meet that test. At Gilbert's death (1603), neither Galileo nor Stevin had yet published anything Copernican, and Kepler was as yet barely known outside specialists in astronomy. Stevin was attacked for sacri-

lege when he published his Copernican book in 1605. Therefore, taking account of context, the simplest explanation of Gilbert's evasiveness is that he judged (or perhaps even had been told) that it would be inappropriate for the queen's personal physician and president of the Royal College of Physicians to be completely candid on such a delicate matter. If you would like a villain to blame for Gilbert's odd combination of repeated positive hints about but always explicit evasions with respect to the Earth's orbital motion, Bacon invites attention. We know that he read Gilbert's books (not only the published *De Magnete*, but in manuscript *De Mundo* as well). Bacon disapproved of Gilbert's way of pursuing science, and also disapproved of Copernicus. And at the time, he was a powerful figure in the Elizabethan court.

Inhibitions on explicit Copernicanism were hardly unusual. Maestlin (Kepler's teacher) was Copernican, but his widely used textbook was not. When Galileo succeeded in getting his protégé Castelli appointed professor of mathematics at Pisa (which, as was usual at the time, included responsibility for teaching astronomy), Castelli was explicitly warned not to teach Copernican astronomy. When he queried Galileo on how he could obey that, Galileo advised him to take the job and teach conventional astronomy, as he himself had done for twenty years—as indeed, throughout at least the first half of the twentieth century, biology was usually taught with no reference to Darwin.

8. It was forty years before anything substantial was done with Darwin's theory, twenty-five years before anything was done with Cournot's pioneering venture in mathematical economics, and several decades before anything striking was done with general relativity. Allowing for the early state of science, the sixty-year delay here is not out of scale with what we could reasonably expect. But we might look for something that accounts for the turning point coming when it did.

CHAPTER 1

Making Worlds

The young Copernicus spent six years in Italy studying law and medicine (and, on the side, astronomy), then returned to Poland to spend the rest of his life in what he called the "darkest corner of the world."[1] But he had an observing tower built on the wall surrounding the cathedral whose domains he helped manage, and for thirty years he worked at his great reformation of astronomy. He did other interesting things as well, although the record is too thin to say much about them. He was called to the court of the king of Poland to give medical advice, and he offered what appears to have been good advice on currency reform to the Polish diet. He was invited to Rome to participate in work on a new calendar.[2] He seems to have organized the defense of the cathedral lands against marauding Teutonic knights. He also got into a bit of trouble through an overly cordial involvement with his housekeeper.

As I've mentioned, Koestler (and others) called him timid. And, no doubt encouraging such appraisals, he began his *De Revolutionibus* with an apology for presenting an idea that would seem "absurd." But *timid* is not the way to characterize what Copernicus was doing here. If you are going to propose something that will sound absurd, you do well to make it clear to readers that you know what you are doing. The Lutheran minister seeing the book through the press sought to make the argument more palatable by assuring readers that the main idea need not be taken seriously. But once the apologies have been gotten out of the way, Copernicus's own opening is emphatically otherwise. By the end of his introduction he is warning philosophers and theologians to be careful about meddling in this business, lest they make fools of themselves.[3]

And despite its "absurd" central point, the book was read and admired, and its author was hailed as "the new Ptolemy." A second edition was printed in 1563, even though the only astronomer who had yet publicly said that Copernicus was right was his sole disciple, Rheticus. The accolades were for Copernicus's technical work, not for his heliocentric idea. For as mathematical systems, the Ptolemaic and Copernican accounts are essentially interchangeable (as you will see in this chapter), so a Ptolemaic astronomer could use Copernican mathematics. On the other hand, even if the book was being studied for what we would see as the wrong reason, Copernicus had put on the table the biggest question that could possibly be raised: What was the true system of the world?

There were three major choices. Ptolemy's enjoyed the advantage of 1400 years of seniority, and the much larger advantage that it put the Earth comfortably at the center of the world. The Copernican alternative of course put the Sun at the center and made the Earth another planet. And a compromise proposed by Tycho Brahe in 1588 allowed the traditional planets to orbit the Sun, but had the Sun (carrying the orbits of the planets) orbiting the Earth.

To the modern eye, a Ptolemaic diagram looks like the innards of a mechanical clock: astronomy as envisioned by a madman. But Ptolemy's astronomy lasted for 1400 years because it works! In fact, as you will see in a moment, if the world is really Copernican, it is *necessarily* true that Ptolemy's astronomy works. For what Copernicus showed was that Ptolemy could be turned inside out (or the reverse) while leaving observations of the planets completely unchanged. And since believing that the Earth is flying through the heavens (as a heliocentric system requires) was not easy, it is hardly puzzling that Ptolemaic (not Copernican) belief came first.

Tracing the Planets

Against the background of fixed stars, the planets can be seen to move completely through the zodiac (the narrow band of stars centered on the annual path of the Sun), returning to about the same position among the fixed stars. But there is a most surprising aspect to these planetary paths: They make backward loops (retrogressions)! Figure 1-1 shows the path of Mars as it moves six times (over twelve years) through its two-year cycle around the heavens.

A Ptolemaic astronomer (or any other kind of astronomer who takes the Earth as the fixed center of the world) would have to say that the planets actually make these looping motions. But a Copernican would say that Mars, as seen from the Earth, which is also in motion, only appears to make occasional backward loops. The loops are an illusion, like the backward motion that telephone poles seem to make against a distant horizon when you speed past them on a country road. Hence how you *interpret* what you see in Figure 1-1 depends on whether or not you see the world as Copernican. But any model that works must generate those loops, since in fact that is what you see.

And if you study the diagram (or the analogous tracing of a simulation in a planetarium shown in Figure 1-2), you'll notice that reproducing such loops theoretically is not an easy task. That the planets make loops at all is puzzling. And the loops never exactly repeat. Over long periods of observation, astronomers can see a stable average period, size, and shape for the looping motions, but from one cycle to the next, the shapes keep changing. The time between retrogressions is never exactly the same, and if you put a ruler against the drawing, you will find that the span of the retrogressions also keeps varying. The end of one cycle never quite merges into the beginning of the next. An astronomer will never see an exact repetition of a sequence of orbits. For the argument here, we will need to consider only the stable average structure. But to get predictions that work, an astronomer would need to account for the variation as well. For a very long time, capturing those motions must have seemed quite hopeless.

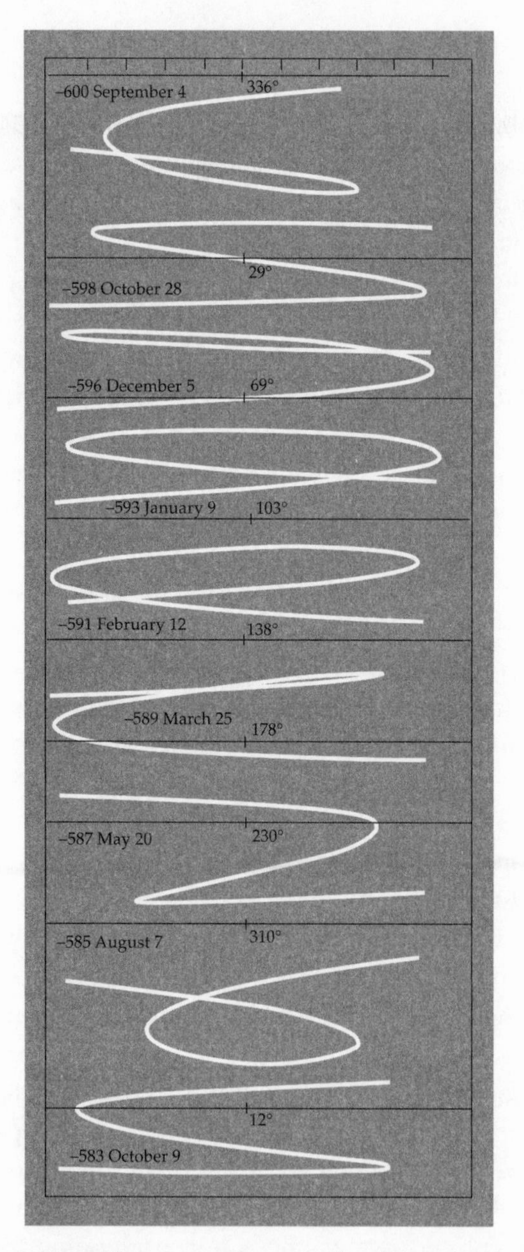

Figure 1-1 Owen Gingerich computed these tracks of Mars showing retrogressions over a dozen years. The paths are backdated to the period in which the epicycle models of planetary motions were first conceived. A contemporary sequence would look essentially the same. (*Reprinted with permission*)

Figure 1-2 This time-lapse photograph of the motion of Mars against a background of fixed stars was simulated in a planetarium. It is another view of loops like those generated by the computer simulation in Figure 1-1. (*Photo by Lessing-Magnum; reprinted with permission of the Deutsches Museum, Munich*)

But Ptolemy did that brilliantly. He was able to work out remarkably simple models—relative to the complexity of the paths themselves—that could generate predictions of these varied shapes and timings. He did it so well that no real improvement in accuracy was made after him until Kepler devised a radically different kind of mathematical model 1500 years later. Copernicus made bold claims for his system, but he did not claim that it was more accurate than Ptolemy's. A few times over the centuries, someone had "reset the clock" with new observations to remove errors that had accumulated since the last resetting, and there was occasional tinkering of various sorts. But in Ptolemy's own version, Ptolemaic astronomy was for all practical purposes as complete as it ever would be . . . or indeed as any astronomical system could be until Tycho Brahe managed a large improvement in naked-eye astronomy in the 1580s.

Double Orbits

Why do the Ptolemaic models work in a world that in fact is Copernican? The key point was noticed long before Ptolemy (c. 200 B.C.)

by the most famous mathematician of the time, Appolonius: A planet moving like Mars in Figure 1-1 or 1-2 could be visualized as being carried on a big wheel that also somehow incorporated a smaller wheel. The big wheel would carry the planet completely around the sphere of fixed stars, and the smaller wheel would provide for the occasional loops. So a workable model might be developed if an astronomer started by giving each planet *two* orbits. This is still a long way from a complete model, since there remained the need for further machinery to handle the cycle-to-cycle variation. But now (about 200 B.C.) we have a start—and a good start, since if for each planet one of the wheels turns to keep pace with the Sun's annual motion relative to the Earth, this builds in the coordination of loops with the Sun/Earth/planet alignment (for Mars, Jupiter, and Saturn) that observations require, and also the closeness to the Sun that observations show for Mercury and Venus.

Today it is easy to see that these relations are inevitable. Parallax effects must occur when the orbiting Earth passes a slower-moving planet or is passed by a faster-moving planet. The changing shape of the loops reflects the changing position of the Earth relative to the plane of the planet's orbit. Planets in orbits bigger than that of the Earth of course must straddle the Earth-Sun distance, and so will sometimes be on the opposite side of the Earth from the Sun. But planets with orbits inside the Earth's orbit can, of course, never do that. Retrogressions for the outer planets can occur only at times when the planet is rising when the Sun is setting. And even a casual awareness of how the planets move shows that two of them (Mercury and Venus) are never far from the Sun.

Figure 1-3 shows three ways to arrange those double orbits. What seems obvious to us, but for a very long time was ignored, was the possibility shown as Figure 1-3*a*. It gives one of the two required orbits to the observer aboard the moving Earth, so that the observed planets themselves do not need double orbits. As a logical possibility, this was *noticed* very early. Just before Appolonius, Aristarchus (c. 225 B.C.), who was the first astronomer to realize that the Sun is vastly larger than the Earth, had proposed that the tiny Earth might orbit the big Sun. But Aristarchus's proposal lay dormant for the next 1800 years.

This was hardly an easy idea to sell. You could think of the Sun (and the rest of the heavens) not as circling the tiny Earth, but as circling the center of the world, where the tiny Earth happened to be. And unless the fixed stars were at a vast distance from the Earth, astronomers would see a shift in the apparent positions of the stars if indeed the Earth were moving around the Sun. But no such parallax could be observed. So Aristarchus's idea was intrinsically hard to believe, its contrary was easily rationalized, and in addition it required a vast empty space between the planets and the stars to make the parallax effect so small that it could not be seen. And until the double-orbit idea was fully developed, it could hardly occur to a person that a heliocentric world could reduce the loops of Figures 1-1 and 1-2 to mere illusions. There is no hint that Aristarchus saw that. On the record, it would be 1700 years before someone would finally notice that one of the two circles required to capture the motions of a planet might be an orbit of the Earth.

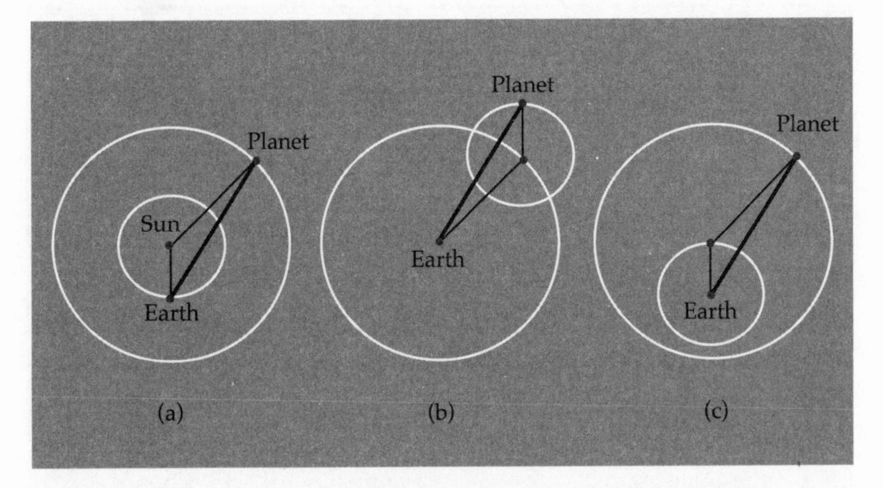

Figure 1-3 Three observationally indistinguishable ways to arrange the two orbits that a planet needs in order to generate both its observed motion around the zodiac and its retrogressions (loops): (*a*) Copernican, (*b*) Ptolemaic, and (*c*) Tychonic.

But Appolonius, who had first seen the two-orbit possibility, did see the equivalence of the little-on-big and big-on-little alternatives (Figures 1-3*b* and *c*). So from the start of what grew into modern astronomy, astronomers knew of two observationally equivalent ways to give planets that were presumed to be circling the Earth their required double orbits: We would say (but until Copernicus no one ever did say) that one circle captures the planet's own motion around the heavens, and the other captures the effect that the Earth's orbit has on observations of the planet's apparent motion. For the Earth's orbit is still there, of course, affecting observations even if the astronomer doesn't realize that it is there.

In Figure 1-3*b*, the smaller orbit (the epicycle) rides on the bigger one, as seems natural. Alternatively, but decidedly odd-looking, the big orbit could be carried by the little, as in Figure 1-3*c*. And as the tint box explaining Figure 1-3 illustrates, a modest exercise in geometry shows the fundamental point that all three possibilities yield exactly the same observations: two that astronomers knew about for a very long time, and a third that was destined to wait in the wings for a very long time.

Equivalence of the Three Double-Orbit Models of Figure 1-3

What eventually became the Copernican setup is on the left (Figure 1-3*a*); what became the standard Ptolemaic model is in the center (Figure 1-3*b*); and an inside-out (inverted) Ptolemaic model is on the right (Figure 1-3*c*). Across the figure, the same pair of circles is combined in three different ways. The Copernican model turns the Ptolemaic model inside out with respect to whether the Sun or the Earth is at the center. But the Ptolemaic setup itself can be turned inside out with respect to whether the big orbit carries the small or the converse. For an outer planet (with a converse argument for the inner planets), in each variant in Figure 1-3 the smaller circle turns with the annual Sun/Earth motion, and the bigger turns with the period of the planet's own motion.

In the Copernican model, the pointers in the two circles obviously turn independently. There is no link between them. Therefore, if the world is actually Copernican, models that take the Earth to be the fixed center of the world must somehow also provide that independence even though now the two orbits are linked: An inner circle carries the center of the outer circle. But given a stipulation that the circles for each planet turn independently (as if riding on frictionless bearings), suppose that God starts the three larger pointers in parallel, and does the same for the three smaller pointers. Since the respective pointers turn at identical rates in all three models, if they start in parallel, they will remain in parallel, which means that the line of observation from the Earth (the dark line in the three diagrams) will remain the same in all three models. So for predicting where a planet will be seen, it makes no difference whatever which model is used.

Complete models would need the additional machinery that handles the variation you see in Figure 1-1. However, the complications would be identical for each of the possibilities. If the three models yield identical predictions for the simplified situation covered here, they will also yield identical predictions when the complications are added.

The Copernican model on the left of course simplifies what is actually happening in the heavens. It eliminates the loops as real motions. But notice that in terms of predicting what will be seen from our moving Earth, there is no simplification at all. What is eliminated from the motion of the observed planet comes back into account for the motion of the observer.

The Physical Proposal

The alternative models of Figure 1-3 yield exactly the same predictions from exactly the same calculations. To say where a planet was, or where it will be, at any particular moment, you need to compute the directions at that moment of the pointer in the bigger circle and the pointer in the smaller circle, treated as turning independently. Together the two pointers determine the line of observation shown by the dark line in each model. It doesn't matter which diagram you use. In practice, astronomers mostly did not use any of the diagrams. They worked from tables that treated the two orbits independently and allowed for the "bisected eccentricity" that Ptolemy had devised to account for the cycle-to-cycle variation.[4] Consequently, if astronomers cared only about correctly predicting the looping motions of the planets, their choice of a model would have had no more scientific significance than their choice of dinnerware.

But actual astronomers (in some contrast here to philosophers philosophizing about astronomy) have almost always cared about what the world was really like. It seems to go against human nature to be satisfied with an incomprehensible formula, even if it works, if an alternative can be found that seems to make sense of why things are as they are. Long before anything that might be called science, human societies always produced myths that explained why the world was the way it was. And Ptolemy, having developed his elaborate mathematical system for predicting the motions of the planets, then produced a detailed account of what he conjectured was physically in the heavens producing those motions. This "planetary hypothesis" became a matter of firmly settled belief among astronomers.[5]

But in this Ptolemy proved to be unlucky. His physical proposal was exceedingly clever, but it was not a good idea. Logically, as illustrated in Figure 1-3, it is easy enough to see that Ptolemy's models could be rearranged as Copernican models. But given Ptolemy's ingeniously worked out physical proposal (to be discussed next), it became especially hard to notice that. It would be hard to believe that anything that works as well as Ptolemy's proposal does could be fundamentally wrong, and in any case it would be hard to believe that the Earth is flying through the heavens. This double burden proved to be insurmountable for a very long time. We need a path-dependent account to make sense of how, after 1400 years, someone

(Copernicus) finally escaped from that formidable box. Ptolemy provided the essential kernel from which the Copernican idea could eventually sprout, but he wrapped that kernel in a hull that proved to be very hard to crack.

Economy and Comfort

It would be a serious mistake to suppose that what Ptolemy did was in any way odd, or that his propensity to make commitments about what lies deeper than he could actually see was some peculiarity of premodern thinking. If that were so, we would have at hand a candidate for what was revolutionary about the way thinking changed in the seventeenth century. But we don't. In our everyday lives, in our political beliefs, and in our science, we inescapably do today what human beings have always done—and it should be emphasized that on balance, we do it with some success. In Jerome Bruner's (1979) phrase, we irresistibly go "beyond the information given." We guess at what is beyond what we can see.

So consider how this propensity worked out in the case at hand. For each planet, observations would tell an astronomer how fast each of its two circles turns. As we've already noticed, one circle tracks the Sun/Earth motion. In a geocentric world, things do not need to be that way. It is a remarkable cosmic coincidence. But an astronomer can see that it is there. And astronomers could see how long it takes each planet to go around the heavens. More subtly, observations can also determine the ratio that must hold between two circles. The average span of a planet's retrogressions is determined by that ratio. Therefore, there is a unique ratio that produces retrogressions that match what the planet does. Ptolemy must have been enthralled to find that with parameters imposed by observations (for the periods of the two circles for each planet and the ratio between the radii of its two circles), his models indeed would correctly capture for each planet the intricately varying, looping motions astronomers saw. There is an understandable sense of ecstasy in the opening paragraphs of his great book. Ptolemy found himself master of a set of abstract models that (within the limits of naked-eye astronomy) solved the problem of the planets. He was tempted to conceive a world in which it would make sense that these *simple* models actually work—simple here meaning relative to the complexity of the observations they can account for and predict.

Ptolemy started from a fundamental intuition that seems to have guided scientific theory building from its origins. The intuition can be phrased in many equivalent ways. Nature loves simplicity. Nature is the best artisan. Nature does not make unnecessary entities. We are after what distinguishes modern science from premodern science, but as with the propensity to build theories that go beyond the information given, we will not find a difference here. There is nothing uniquely modern about this propensity to suppose that Nature prefers simplicity, elegance, and in general what looks to us to be good design as against what looks ugly, clumsy, redundant, and so on. To give this a label, call it the propensity to favor *economy*.

Thomas Kuhn perhaps misled readers when he called this kind of preference *aesthetic*, which (to Kuhn's dismay[6]) many readers took to be just a synonym for "arbitrary" or even "irrational." Theory preference can't be reduced to a formal calculus. You can't even *prove* that the Sun will rise tomorrow. But belief is not arbitrary. Musicians have no trouble agreeing that Mozart's music is better than Salieri's. This is not a preference that might plausibly be reversed next year, like the preference for wide versus narrow neckties. The Mozart-Salieri preference is not demanded by a formal calculus, but music lovers still find the choice inevitable. The contrast between scientific theories that win and those that lose is usually at least as severe as the contrast musicians perceive between Mozart and Salieri, and often a great deal more severe. Some theories make better sense of the world, and they are hard to beat.

There is another aspect of theory choice that might reinforce—but that sometimes competes against—the taste for economy. We prefer what feels comfortable: what is convenient to use, what fits easily with our habitual movements or (unconsciously, but particularly germane here) with our habitual patterns of thought. When choosing between two theories that seem equally economical, we will always find ourselves preferring the one that fits more comfortably with other cognitive patterns in our repertoire. Similarly, between theories that are equally comfortable, we will always prefer the more economical. And we don't *choose* to think that way. It is the way our brains work that chooses that way.

In concrete cases the choice may not be so simple, since what seems comfortable and what seems economical will change over time.[7] What seems at first to be uncomfortable will become less so if it works so nicely that we are led to make use of it anyway. We become accustomed to its oddities. What seems complicated when we are new to it comes to seem simpler as we become familiar with its arguments. And this is all the more true when those arguments are polished, and especially if they take hold with people we have confidence in, or with people in general. Even a terribly difficult argument can come to be reduced to a one-step argument: "Everyone knows" it's right. When something takes hold as social knowledge, it begins to seem automatically both more economical and more comfortable. Sometimes logically, sometimes merely by trained inattention, we learn to ignore aspects of it that might otherwise seem clumsy.[8]

Ptolemy's System

Ptolemy's system exhibits all these features relating to comfort and economy, including trained inattention, since Ptolemaic astronomers learned to mostly ignore the system's anomalies.[9] Figure 1-4*a* shows Ptolemy's physical arrangement for a single planet, as Copernicus and his contemporaries would have known it.[10] This "sphere" moves like the platform of a merry-go-round, turning in place without disturbing either what is in the center or what is outside.

For an actual merry-go-round, the motive power is provided by machinery that occupies the central space. But Ptolemy's heavenly spheres needed no such machinery. Ptolemy's spheres are eternal objects, possessing by their nature an intrinsic rotation. This may

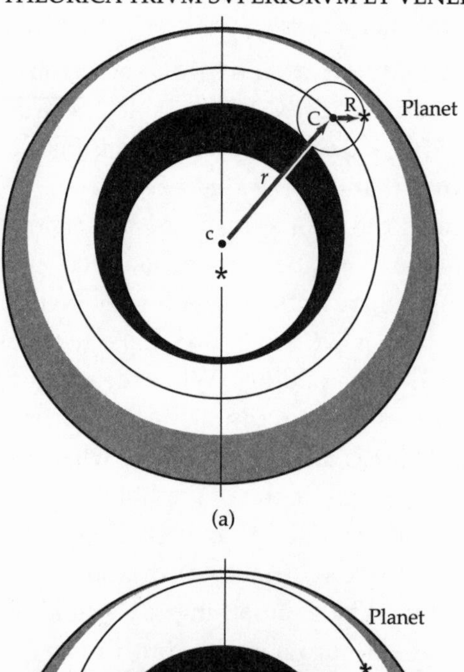

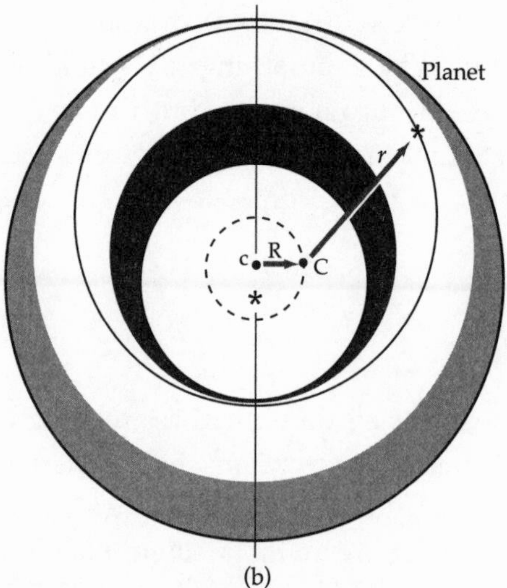

Figure 1-4 (*a*) This diagram is adapted from Peurbach's hugely influential *New Theory of the Planets* (some sixty editions after its first printing in 1474), which introduced Ptolemy's account of how to physically interpret abstract schematics of the sort shown as Figure 1-3*b*. (*b*) In this variant, the large and small orbits have been interchanged: the large orbit now rides on the small, as illustrated in Figure 1-3*c*. Observations generated by both alternatives will be identical. (*Courtesy of the University of Chicago Library*)

sound quaint, but it is not so different from the initial push that Newton imagined that God gave the planets, which was sufficient to keep them eternally going around their orbits even as you read this.

For Newton, the eternal motion was in a straight line, and the orbits were created by continuous gravitational deflection from that straight-line tendency. The Sun's gravity reaches invisibly (and, let's face it, mysteriously) through space. But for Ptolemy, the natural motion of heavenly bodies was circular. We think Newton got it more nearly right, and we have good reason to think so. But given what was known to Ptolemy, or to Galileo only a few decades before Newton, circular inertia around a center of gravity was a reasonable account of what they could see in the world.

In Ptolemy's scheme, each planet is located within a transparent, perfect sphere, and these spheres fill the heavens in a nested arrangement, with Earth at the center. There is no displacement in the heavens at all. A Ptolemaic sphere rotates in place relative to whatever is in the center. Within the sphere of Figure 1-4*a*, you can see a second object, which also rotates in place relative to the sphere that contains it. So the sphere rotates in place relative to its center, and within it an *epicycle* (which means the cycle displaced from the center) rotates in place relative to the sphere. And on the perimeter of the epicycle is the actual planet. It has no motion at all of its own; it just sits in its place on the epicycle, which turns in place within the sphere, which turns in place around the center of the world, where sits the Earth.

The combined rotation in place of the whole sphere and the rotation in place of the epicycle within the sphere produces the planet's looping motions. With the help of Ptolemy's bits of secondary machinery, the planet correctly traces out the complicated paths illustrated by the diagram of Figure 1-1. You could think of the epicycle in Figure 1-4*a* as a circular track mounted on the merry-go-round platform, with the planet as a toy train chugging around the track as the platform carries it around the center. If the train carried a light on its engine and you ran the merry-go-round in darkness, the light you saw would trace out looping motions like those that actual planets trace out. But for Ptolemy, it is the whole track carrying the planet that rotates as it is itself simultaneously carried around within the sphere.

From the argument reviewed in connection with Figure 1-3, exactly the same motion generated by the sphere of Figure 1-4*a* would also be provided by an inverted version of that sphere, as in Figure 1-4*b*. In Figure 1-4*b*, the epicycle (the "track") is the big circle and surrounds the central region. The radius from the center of the sphere to the center of the epicycle (marked by the asterisk) is exactly the same as the radius of the epicycle in the little-on-big setup (Figure 1-4*a*). Similarly, the large circle in Figure 1-4*b* (now carried on the outside as the epicycle) is exactly the same size as the large circle that carries the epicycle in Figure 1-4*a*. And since the rates at which the corresponding rotations in each figure move stay the same and since the size ratio is the same, the path generated for the planet in Figure 1-4*b* will be identical to the path for the planet in Figure 1-4*a*.[11]

In the *Almagest*, Ptolemy mentions the big-on-little inverted[12] model as a possibility for the outer planets. But then he promptly moves ahead with an exposition that uses only the more comfortable small-on-big diagrams. As any reader will notice in trying to envision the motions of the alternatives in Figure 1-3, little-carried-by-big motions are intuitively comfortable, but big-on-little motions are hard to make sense of, given our experience with how things move in the world. By the time Ptolemy came to construct his proposal for a physical embodiment of those abstract models, the inverted possibility seemed to have been forgotten. So although this possibility was noticed, it remained dormant, like Aristarchus's heliocentric idea, for another 1400 years.

The Nested Spheres

Ptolemy's response to the economy dimension (of the comfort/economy pair) shows itself in an axiom that guided his world building. He specified that the heavenly machinery should have neither overlaps nor empty spaces. These stipulations satisfied two goals at once, which is prima facie the economical way to proceed. There was no need to explain how overlap occurred without confusion among the planets, and there was no need to consider what purpose the empty space might serve. And given this no-overlaps/no-gaps condition, the size of all the orbits was also determined.

Ptolemy was free to scale his individual spheres any way he pleased. Changing the scaling (enlarging the pair of circles for one planet, shrinking them for another, and so on) would not affect the angle of observation shown in Figure 1-3. Together with his principle of no overlaps and no gaps, this suggested a particularly plausible arrangement. Ptolemy set the inner edge of each sphere to come right up to the outer edge of the sphere just inside it. For each planet, once the distance (from the Earth, of course, for Ptolemy) to the inner perimeter of a sphere was set, that distance uniquely determines how big its epicycle must be if it is to yield retrogressions that will match observations. Only one value will work. In turn, that epicycle determines the distance to the outer perimeter for that planet's sphere, and hence also determines the inner perimeter of the next sphere. When Ptolemy ran out of spheres for the planets, he put an outermost sphere carrying the fixed stars, leaving to the gods anything beyond that.

Ptolemy also had a way to choose the order of the spheres. Of the heavenly objects visible at the time (that is, to the naked eye), only the Moon was close enough to allow observations of parallax accurate enough to determine its distance. But given that Earth-Moon distance, the location of the inner edge of the first planet became the minimum distance that would leave room for the motions of the Moon. From there, successive spheres were snugly nested until the set was complete and enclosed by the sphere for the fixed stars. The space between the Moon and the fixed stars was completely filled by nested spheres like that of Figure 1-4*a*. There were six such spheres: one for each of the five visible planets and the sixth for the Sun, each with a defined maximum and minimum distance from the Earth.

Even the Sun rode on an epicycle, although this epicycle was too small to be shown in diagrams like those from Kepler and Tycho, to be introduced next. The Sun's epicycle was not big enough to produce loops, but it did account for the Sun's acceleration during northern winters (and its slowing during northern summers), which our calendar reflects by making February several days shorter than August. The point at the center of the Sun's epicycle was called the mean Sun.[13]

Fitting the Planets

Indisputably, the Moon is closest to the Earth, since it eclipses all other heavenly bodies. The Moon also has (by far) the most complicated motions of all the heavenly bodies. In the Ptolemaic view, it was intuitive that the closest body would also exhibit the most complex movements. Today we would say that the Moon's orbit is perturbed by three-body gravitational interactions with the Sun and the Earth; but for Ptolemy, the complications were due to the Moon's proximity to the corruptible sublunary world.

Of all the bodies beyond the Moon, Mercury has the most erratic movements. It is the only naked-eye planet, we can say today, whose orbit is elliptical enough to be easily distinguished from a circle. In the Ptolemaic argument, the complicated movements of Mercury (compared to those of the other planets) show that it too is close enough to be vulnerable to sublunary influence.

Next must come Venus, which, like Mercury, is always seen near the Sun, but whose orbit is almost perfectly circular. Then comes the Sun, dividing the two classes of planets (those that are always seen near the Sun and those that orbit independently of the Sun). From its central location, the Sun can give each planet a component of motion tied exactly to its own annual motion. Finally, Mars, Jupiter, and Saturn follow in the common-sense order suggested by their increasingly long periods of revolution (two, twelve, and thirty years, respectively).

Working back from the sphere of fixed stars, we would then have Saturn snugly nested just inside that outermost sphere. Jupiter would snugly fit in the open space inside Saturn, then Mars inside Jupiter, and continuing with the Sun, Venus, and Mercury, the Moon, and finally, at the center of everything, the Earth. The entire ensemble would be neatly fitted like a set of Russian dolls.

The Grand Coincidence

For a Ptolemaic astronomer, there was then a very striking confirmation of this ordering, and indeed of the whole Ptolemaic system. Even though the Sun was too distant from the Earth to allow measurement of parallax, there were ways in Ptolemy's time to estimate its distance relative to the distance to the Moon. From observations of the shadow of the Earth or the Moon at eclipses, Ptolemy calculated an Earth-Sun distance. This was taken to be an observable matter of fact, even though a certain amount of determined negligence was needed to fail to notice that the accuracy required to actually make Ptolemy's calculation convincing could never be achieved with naked-eye observations. After Ptolemy this Earth-Sun distance settled into being treated as well established, subject from then on only to slight adjustments. The number was never actually justified, but neither was it ever seriously doubted.[14]

If that Earth-Sun distance was taken as known—a step that was never questioned through all those centuries—there was a very severe constraint facing anyone who wanted to assemble a physically credible scheme of nested spheres. The inner planets would have to be such that their combined ratios of greatest-to-least distance from the Earth would just fill (with no overlaps and no gaps) the available space between the nearest approach of the Sun and the furthest distance of the Moon. So there had to be just the right amount of room for the epicycles of whatever planets came between the Moon and the Sun. Since the individual ratios of the two circles for each planet were determined from observations, Ptolemy could not arbitrarily make the planets fit: They had to just turn out to fit.

But all these criteria are met! The minimum space required for Mercury and Venus is, most remarkably, almost exactly what is needed to allow those two planets to neatly fill the available space. The orbits of the very planets that would most plausibly be fitted between the Moon and the Sun (Venus and Mercury) can be calculated from observations to require a space that matches the space that calculations make available between the Sun and the Moon. As Neugebauer, the most revered authority on these matters, comments: "A most implausible accident . . . became the solid foundation of the [nested-spheres system] which dominates medieval astronomy."[15]

The symmetry that then appears in the Ptolemaic arrangement, with the Sun between the two kinds of planets and with the epicycles diminishing in relative size as you move away from the Sun in either direction, becomes yet another confirmation of the divine ordering imposed by the nested-spheres logic. This elegant system then provides a very detailed and concrete picture of the world, of distances to the various planets and to the fixed stars, and even seemed to determine the sizes of the various planets.[16] And this machinery predicts how the planets will move with remarkable precision.

Figure 1-5 is Kepler's illustration of this Ptolemaic world, intended of course (coming from Kepler in 1596) to be contrasted unfavorably with the Copernican world. It shows a slice of Ptolemy's system near conjunction of the three outer planets, so that all the epicycles are in line, providing another look at the "Russian dolls" setup described in connection with Figure 1-4*a*. If you could show it to a Ptolemaic astronomer in 1543, when Copernicus's book appeared, this drawing of Kepler's would look quite beautiful, while Copernicus's drawing of a world with the Earth flying through space would look simpler (an astronomer might allow), but quite crazy.

But now a letdown. I have described this Ptolemaic world with some enthusiasm, hoping to convey a sense of its immense appeal. It endured for fourteen centuries. But the neat fit was, in fact, a delusion. Although the Earth-Sun distance was solidly entrenched as something that "everyone knows," it was really only a highly arbitrary choice of a highly agreeable number. Given the imprecision of naked-eye observations, the fit was entirely meaningless. But the number was so thoroughly entrenched that even Copernicus did not

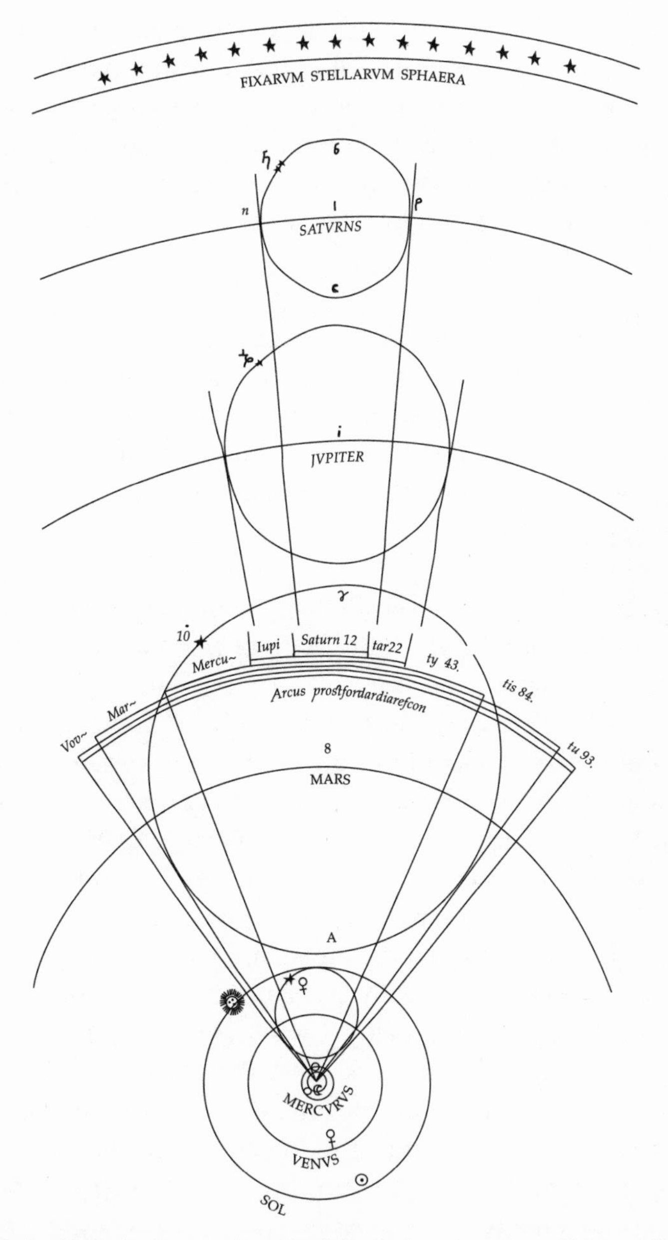

Figure 1-5 Kepler's diagram (1596) of the Ptolemaic system. The arcs Kepler has drawn show the epicycle-to-deferent ratios of Ptolemy's outer planets and the inverse for the inner planets; these ratios translate easily into ratios of heliocentric distances for Copernican planets. (*Courtesy of the University of Chicago Library*)

do more than mildly quibble with it. No Ptolemaic astronomer saw the number as problematical at all. What to us is transparently circular reasoning seemed for a long time to be transparently correct reasoning. A highly motivated Kepler finally suggested that the number was probably off by a factor of 3. Eventually astronomers realized that it was actually off by a factor of 20.[17]

Inverting Ptolemy

But there were other ways to construct a world that puts flesh on the bare-bones possibilities of Figure 1-3. How far each of these possible worlds was even noticed, and how far each (if noticed) attracted supporters, must tell us something about how people who concerned themselves with such issues were seeing the world in the years before 1600.

An astronomer could invert Ptolemy's outer planets and produce the alternative world in which the paths of Mars, Jupiter, and Saturn would be generated by the motions in Figure 1-4*b* rather than those in Figure 1-4*a*. This yields an alternative Russian-doll system that is identical to the standard Ptolemaic system out to the orbit of the Sun, but in which beyond that the planets have the inverted (big-on-small) motions of Figure 1-4*b*. Call this *inverted* Ptolemy. The paths actually followed remain identical in both setups; and the thickness of each planet's sphere (from inner to outer boundary) is also identical. The neat Ptolemaic fit is not disturbed at all. Logically, this inverted version of Ptolemy was a readily available alternative, since Ptolemy himself mentions it as a possibility. There was no need for a discovery.

And although big-on-little double orbits taken one by one are less intuitive than little-on-big, inverted Ptolemy has one big advantage as a *system*: It aligns the first motion of *all* the planets (not just Mercury and Venus) with the Sun. So the entire space from just beyond the Moon out to the fixed stars can move as a single sphere, carrying within it both the Sun and the individual orbits of all the planets in a common annual motion. So unlike Ptolemy's usual version, the inverted system provides an explanation of *why* each of the planets has a component of motion that moves exactly with the Sun's motion. As in the Copernican setup on the left of Figure 1-3, but for a completely different reason, that is automatic, given the inverted arrangement of the outer planets.

The difference between the inner planets (Venus and Mercury) and the outer planets (Mars, Jupiter, and Saturn) now becomes just a matter of the size of the planet's own orbit. The two planets with orbits smaller than the Sun's continue to fit neatly between the Earth and the Sun. The planets whose orbits are bigger than the Sun's straddle the Earth-Sun, so they are sometimes on the opposite side of the Earth from the Sun and sometimes on the same side. In this inverted system, all the planets fall into a single natural order according to the size of their individual orbits, which is also an ordering (now including the Sun and Moon) in terms of increasing period.[18]

Between the standard and inverted Ptolemaic setups, neither arrangement could claim an *absolute* advantage. For example, big-on-little orbits require what appears to be backward rotation of the epicycles of the outer planets. And a geocentric astronomer could prefer the symmetries I've described of the standard arrangement, as the record shows that Ptolemy, and for a very long time astronomers in general obviously did. But—and this is the story we need to consider in Chapter 2—between 1543 and 1600 something happened that changed this preference among astronomers. After 1400 years, the inverted arrangement finally took hold, but in a variant that surely would have surprised Ptolemy.

Notes

1. From his dedicatory letter to the Pope in *De Revolutionibus*.
2. Copernicus declined the invitation to Rome, perhaps for his stated reason (that the effort was premature), but perhaps because he could not get approval for yet another prolonged stay in Italy.
3. *De Revolutionibus*, trans. Duncan 1976, pp. 26–27: "There may be triflers who though wholly ignorant of mathematics nevertheless abrogate to themselves the right to make judgments about it because of some passage in Scripture wrongly twisted to their purpose, and will

dare to criticize and censure this undertaking of mine. I waste no time on them, and indeed I despise their judgment as thoughtless. . . . Mathematics is written for mathematicians, to whom this work of mine, if my judgment does not deceive me, will seem to be of value to the ecclesiastical commonwealth over which your Holiness now holds dominion."

4. Gingerich (1993, pp. 174–175) gives a particularly clear account that includes a simple geometrical demonstration of how Ptolemy's device approximates the effect of Kepler's second law. I give an account of Ptolemy's system in Chapter 7 of my 1993 book.
5. See Van Helden, 1985, ch. 4.
6. See his afterword in the second edition of his famous book on scientific revolutions (Kuhn, 1971).
7. Margolis, 1993, pp. 147–160.
8. My 1993 book is an extended effort to provide an account of the problem of getting over the (cognitive) barrier that separates a revolutionary idea from what it might eventually displace. We must expect that during such a transition there will be a few people who come to see the novelty as too pretty to be wrong, while others are seeing it as too ugly to be right. It is not that aesthetics drive one preference but not the other, but rather that the taste for economy and comfort can weigh out differently for different people, even when both are competent, well intentioned, etc. Such effects also occur (of course) when critics are confronted with radical novelties in the arts.
9. See Margolis, 1993, Chapter 7. The principal bit of Ptolemy's secondary machinery violates the notion of unchanging circular motions in the heavens. Occasionally Ptolemaic astronomers were bothered by that, and in fact it seems to have been crucial in provoking the inquiries that led Copernicus to his great discovery. Ptolemaic planets vary in speed continually as they are carried around their orbits, which is good for Ptolemy's ability to predict how the planets will move (since in fact that is how planets move), but inconsistent with his stated commitment to simple, uniform motion. The planet's period is fixed, and even the rate at which it moves on any segment of the orbit is fixed. But that rate changes from segment to segment. An equant point would be offset above the center of the sphere in Figure 1-4*a*. It is called equant because it is from this point, not from the actual center, that an observer would see uniform motion. The actual center of the sphere, in turn, is offset symmetrically from the location of the Earth (marked "mundi" in the diagram), yielding Ptolemy's "bisected eccentricity." This setup looks very much like a nearly circular ellipse with the central

object at one focus of that ellipse, which hints at why Ptolemy's models turn out to work so well.

10. This sphere will work for any outer planet (Mars, Jupiter, or Saturn), and for Venus as well. Because of its markedly elliptical orbit, Mercury requires an extra epicycle. For the outer planets, with orbits bigger than the Earth's orbit, the epicycle carries the planet while rotating parallel to the Earth-Sun motion; the deferent (the white annulus within the sphere) turns with the planet's own period. For Venus and Mercury (but with the secondary epicycle for Mercury), the reverse holds.

See North (1995, p. 255) on Peurbach's influence, and for a particulary good survey of the whole subject.

11. Departing from usual usage, I will always refer to the circle being carried as the epicycle, even when it is the inverted larger circle. The latter would usually be called a moving eccentric, but it makes sense to avoid adding another technical term here.

12. In allowing the inverted form (Figure 1-3*c*) only for the outer planets, Ptolemy suggests a concern for physical plausibility, not just mathematical adequacy. Merely as a geometrical exercise, it is as easy to invert the inner as the outer planets. But there is no plausible physical interpretation of that, in contrast to the immediately interesting interpretation of a system with inverted outer planets that we will come to.

13. I will follow Kepler's example in leaving implicit (in this kind of discussion) the distinction between the Sun proper (on its epicycle) and the mean Sun (a kind of average motion of the Sun).

14. For a technical discussion, see Van Helden, 1985.

15. Neugebauer, 1982, pp. 111–112.

16. Astronomers estimated the apparent size of the various planets from their brightness. Combining the apparent size with the calculated distances of the Ptolemaic spheres yielded definite sizes for each planet. Whether anyone would have taken such estimates as reliable given an alternative is doubtful. But for many centuries there was no challenge from an alternative. Tycho (see Chapter 2) extended that procedure into an argument that if Copernicus were right, the stars would have to be as huge as the entire orbit of the Earth. But here, Copernicans (who were highly motivated to challenge such estimates) reminded their adversaries that a distant light seen against a dark background looks huge compared to its actual size. A light on the deck of a cabin seen across a lake can look as large as the entire cabin in daylight.

17. See Van Helden, 1985.

18. The period of the Sun's own epicycle, mentioned earlier, is exactly one year.

CHAPTER 2

Tycho's Illusion

The most fashionable view of the Scientific Revolution (as this is written) is that in fact there was no such thing, that there was no particular turning point that marks the beginning of the kind of science we take for granted today.[1] However, if you look back at Table I-1, it surely appears that indeed "something happened" in the years immediately around 1600. On the argument to come, the reason it has proved so elusive is that what changed was not something explicit, but a habit of mind. What I am going to call *around-the-corner inquiry* made its appearance.

But showing a novel habit of mind is not easy. We cannot directly see it. Even indirectly, it will be hard to notice if it has taken hold so thoroughly that we now all share it. The habit then has to be teased out of the taken-for-granted background. If something changed inside people's heads, it wouldn't turn up in their urine.

And even if we identify the emergence of a novel habit of mind, it may seem insufficient to account for the large shift exhibited by Table I-1. The key notion of around-the-corner (as opposed to direct) inquiry may seem too elusive to be pinned down in a useful or convincing way. Or it may be hard to believe that something that seems obvious to us could actually have emerged only around 1600. If the change was merely that people started doing something that (to us) seems to be just what sensible people would have been doing all along, we would also want to say why the change finally occurred when it did. The argument to come is intended to respond to all these entirely reasonable concerns.

It turns out that the "why *then*" question allows for an unexpectedly clear answer, which in turn helps with the more fundamental "what changed" question. Showing a temporal marker is the business of this chapter, in which an episode that is familiar to historians of science turns out to be far more odd than has been noticed.

The Ptolemaic system dominated astronomy for 1400 years. Then, abruptly, it simply faded away—not in popular awareness, of course, but within the network of people who were actively interested in astronomy.[2] This transition occurred half a century after Copernicus had published his alternative system in 1543, but nevertheless when there were still few Copernicans. Introductory texts would remain Ptolemaic for another generation. But by the end of the sixteenth century, it is hard to find an active defender of Ptolemaic astronomy. Beyond the level of elementary texts, the Ptolemaic system appears to have been mostly replaced by a new variant of geocentric astronomy introduced by Tycho Brahe in 1588.

As on a number of other important points, this goes against what has almost always been said about the Tychonic episode. But the evidence you will see reviewed here is strong. T. S. Eliot wrote of a world that ended not with a bang but a whimper. But the Ptolemaic world seems to have faded away, and quickly, without *even* a whimper.

The Lord of Uraniborg

When Tycho Brahe proposed his new system, he was already the most famous living astronomer. He had a silver nose (a by-product of a student duel) and a personal fiefdom on an island off the coast of Denmark (courtesy of the king). Here Tycho[3] built his "Uraniborg" ("the heavenly castle"), from which he and his staff of assistants observed the heavens with the largest and most exact astronomical instruments ever constructed. He could publish his results in large, handsome books from his private press. In the fields surrounding Tycho's castle, the island's peasants toiled to pay for all this.[4]

The grandest product of all this activity was a new system of the world. Figure 2-1 is Tycho's diagram. His planets go around the Sun, just as in Copernicus's system. But the entire ensemble then goes around the Earth, which remains at the center of the world, just as in Ptolemy's system. This system, said Tycho, dispensed with both "Copernican absurdity" and "Ptolemaic redundancies." And his view has been widely shared.

In Ptolemy's system, a component of each planet's motion exactly tracks the apparent motion of the Sun around the Earth. For Venus or Mercury, the inner orbit (recall Figure 1-3) moves directly with the Sun, and the outer carries the planet's own period. But for Mars, Jupiter, and Saturn, the reverse holds. Before Copernicus, these separate connections to each planet could seem to be just the way an inscrutable God made the world. The Ptolemaic arrangement that exhibits these connections has, after all, a certain elegance. (I tried to show that in Chapter 1.) Absent an alternative, these Ptolemaic symmetries could come to look convincingly right. Ptolemy himself and his followers for more than a thousand years certainly saw it that way.

But Copernicus provided a connection between the Sun's apparent motion and the motion of the planets that was transparent and automatic. The annual component of each Copernican planet's observed motion merely reflects the Earth's own annual motion. The first advantage of Tycho's system was that it would also explain (rather than merely accept) the link between the motion of the Sun and that of the planets, and without the Copernican absurdity of supposing that the Earth flies through space. Since the Tychonic orbits are carried along with the Sun in its annual motion, of course each planet will have an annual component in its motion.

And Tycho could also explain why all previous astronomers had missed this possibility. Because of the intersection of the paths of the Sun and Mars that you see in Tycho's diagram, it seemed that Tycho's system could not work in a Ptolemaic world, where solid spheres carry the planets. But Tycho said he had discovered a proof

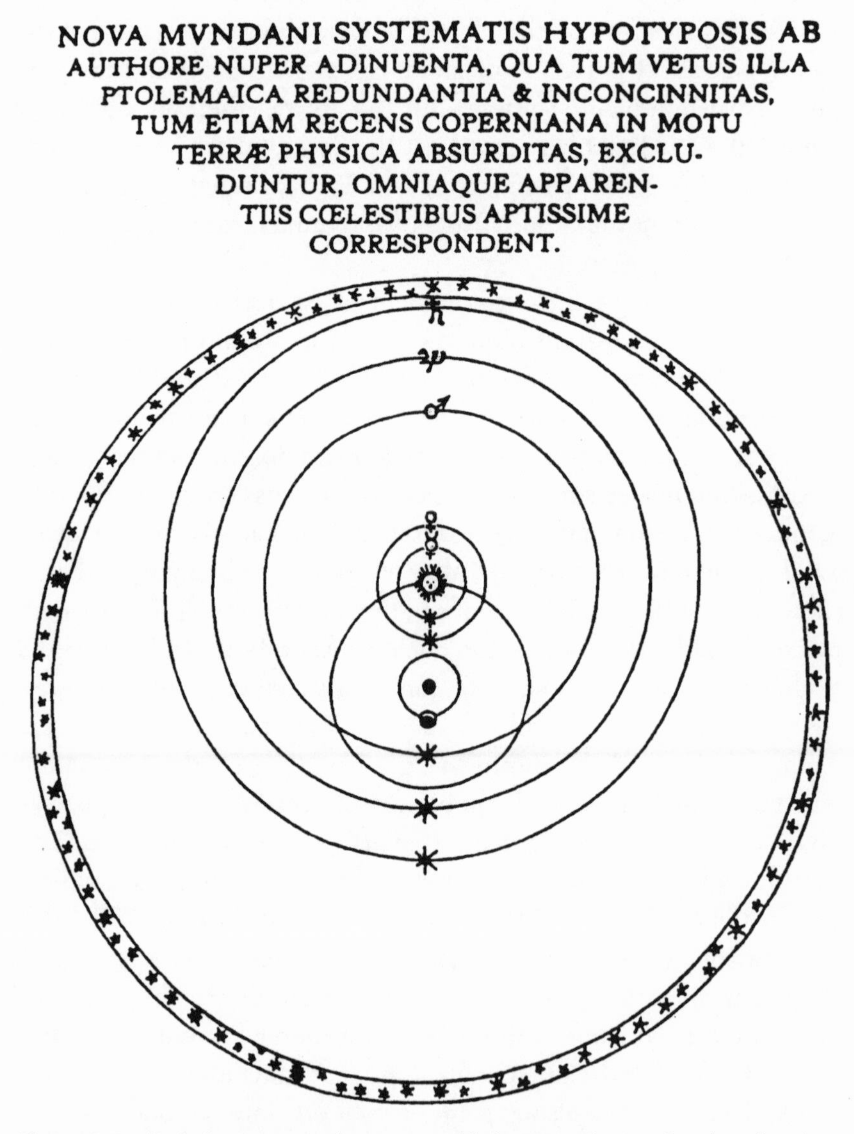

Figure 2-1 Tycho's system, as he presented it in 1588. The caption describes it as "A new system of the world . . . without the Copernican absurdity or the Ptolemaic redundancies." (*Courtesy of the University of Chicago Library*)

that there are no solid spheres, which set the stage for a very neat story.[5] In this neat story, Tycho's system is taken to be intrinsically more plausible than either Ptolemy's or Copernicus's. But in Ptolemy's world of solid spheres the Tychonic system would be (Tycho said) physically impossible. The spheres of Mars and the Sun would collide. But then Tycho's studies of the comet of 1577 showed that in fact there were no heavenly spheres, clearing the way for the superior system. Now Tycho could provide an alternative that left the Earth where philosophers and theologians and anyone with common sense knew it should be, but nevertheless provided "a manifest cause . . . why the simple motion of the sun is necessarily involved in the motions of all five planets."[6]

Furthermore, Ptolemaic models for individual planets could be enlarged or shrunk in any arbitrary way, leaving the actual arrangement of the heavens in doubt (recall the discussion of that point in connection with Figure 1-3). Ptolemy could offer the plausible defense of his ordering that was reviewed in Chapter 1. But in Tycho's system, since the planets all share very nearly the same center for their annual motion, there is no need for conjecture at all. In the Tychonic world, the spacing of the planets is automatically determined.[7]

So we have a most considerable bundle of virtues. With the imaginary solid spheres erased, the way is open for the Tychonic system. This system provides an automatic link between the Sun and the annual component of motion that each planet displays. And as a bonus, the order of the planets is definitely established.

Given all this, why would anyone want to accept the Copernican alternative, which can match these virtues, but at the cost of tossing the Earth into the heavens? Tycho's most dedicated biographer remarks that "until the advent of the telescope, at the very earliest, the available evidence did not render belief in the mobility of the Earth even plausible, let alone convincing."[8] If that were really so, it would be a mystery why all four men who proved themselves to be the ablest discoverers since Ptolemy were already Copernican *before* the telescope was invented. But Thoren's view is the usual view. Thomas Kuhn's widely shared assessment was that with Tycho's proposal, "all the main arguments against Copernicus' proposal vanish . . . and this reconciliation is effected without sacrificing any of Copernicus' major mathematical harmonies."[9]

But there is a profound flaw in this usual story, and it is a flaw that provides an important clue to what could have prompted the eruption of discovery c. 1600. For as will be seen here, although the main arguments *against* Copernicus do indeed vanish, the crucial argument *for* Copernicus also vanishes, leaving a serious question of how anyone could suppose that Tycho offered an *attractive* compromise. What will emerge here is that the neat Tychonic story unravels from both ends.

The Abrupt Decline of Ptolemy

Tycho proposed his system by tacking several pages onto the back of a big book devoted to the comet of 1577, the great bulk of which had already been printed. Apparently he was concerned that someone else would publish first. He said that he had been thinking about the system since 1583, but that he had been held back because of the intersection of the paths of the Sun and Mars that you can see in Figure 2-1. But (Tycho said) the comet of 1577 removed that difficulty, since it showed that solid spheres did not exist after all. The planets simply swam through space, "divinely guided under a given law."[10]

Details on all this were to be provided in a supplement to his book.[11] But Tycho never produced either the promised details on his system or the promised argument showing why there were no solid spheres. Nor did anyone else. Nor, in fact (contrary to the usual claims), was any really good argument available.[12] And although it would be 400 years before anyone noticed, it eventually also turned out that Tycho was mistaken in claiming that his system is incompatible with a world of solid spheres. That claim turns out to be only a kind of optical illusion.

Tycho's error on this point can be demonstrated so simply (see Figure 2-2 and an online animated representation[13]) that it is hardly

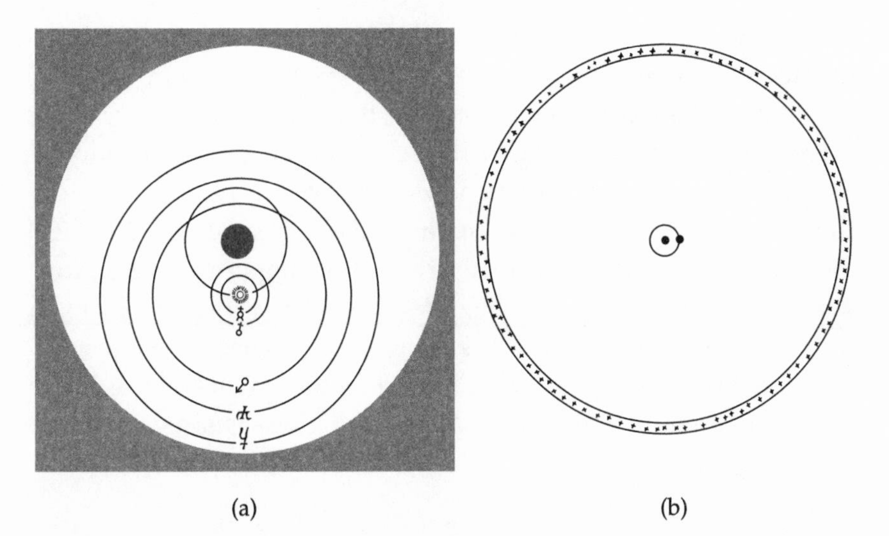

(a) (b)

Figure 2-2 Tycho (and, for the next 400 years, everyone else who commented on the matter) said that his system could not work in a world of solid spheres, since the paths of Mars and the Sun collide. But although the planetary orbits would be solid objects in such a world, carrying the planets around the Sun, the Sun itself needed no such solid orbit. Its path was like the equator on a map: you can draw it onto the paper, but it is not a solid object you can collide with. To see that, make an enlarged copy of the figure, which is simply Figure 2-1 divided into two components. Cut away the black areas around the perimeter and at the center of the panel on the left. You now have a solid model of the Tychonic planetary system, accurate even to roughly the appropriate thickness. Centering this on the template on the right, you can see how an annual rotation of this disk, combined with rotation of the individual planetary orbits around the Sun, produces the correct motions for both the Sun and planets with no risk of collision. (*Reprinted by permission from my report in Nature, vol. 392, p. 857, ©1998, Macmillan Magazines Ltd*)

possible that all Tycho's contemporaries could have missed it, unless by this time no one competent to deal with the issue was motivated to publicly defend the spheres. On the vastly larger matter of Tycho's denunciation of Ptolemy's system of the world, nothing seems to survive beyond controversy over who deserves credit for this move.

The last substantial contribution to *Ptolemaic* astronomy came from a sometime rival of Galileo, Giovanni Magini.[14] In 1589, just about the time Tycho's comet book with its tacked-on new system of the world was reaching its first readers, Magini published a book updating Ptolemy by incorporating material (but not, of course, the moving Earth) from Copernicus. But Magini would write no more Ptolemaic books, since he then almost immediately abandoned Ptolemy for Tycho. Tycho had an assistant visit Magini, and before the end of 1590, Magini had responded with his support for Tycho's innovation. Eventually Tycho advertised this support to every astronomer in Europe by reproducing Magini's letter (see Figure 2-3) in a handsome folio of engravings of his famous astronomical instruments.[15]

At the textbook level, Tycho replaced Ptolemy only in the 1620s (as more recently elementary biology texts long ignored Darwin).

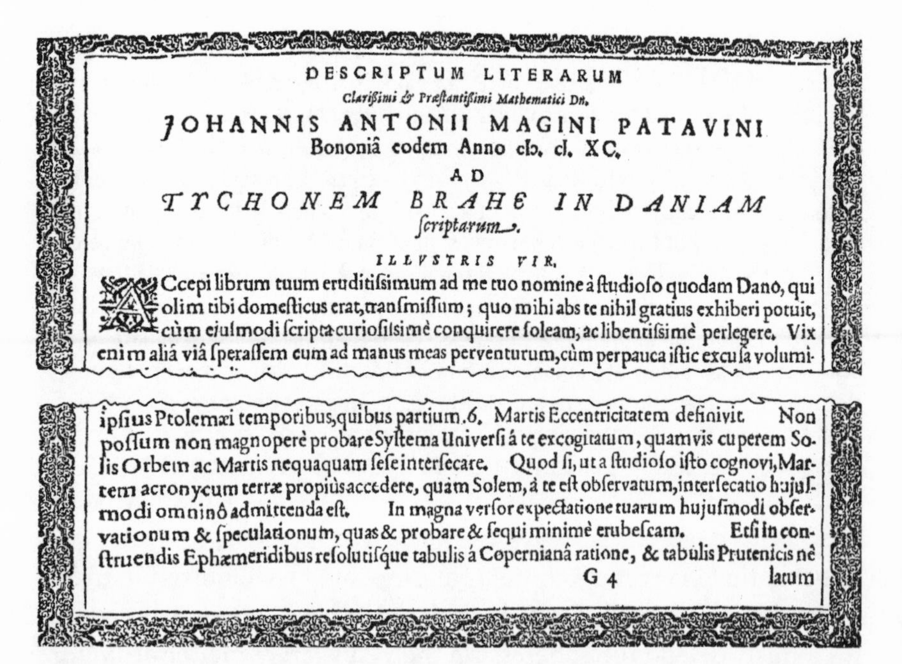

DESCRIPTUM LITERARUM
Clarissimi & Præstantissimi Mathematici Dn.
JOHANNIS ANTONII MAGINI PATAVINI
Bononiâ eodem Anno cIɔ. cI. XC.
AD
TYCHONEM BRAHE IN DANIAM
scriptarum.

ILLVSTRIS VIR.

Accepi librum tuum eruditissimum ad me tuo nomine à studioso quodam Dano, qui olim tibi domesticus erat, transmissum; quo mihi abs te nihil gratius exhiberi potuit, cùm ejusmodi scripta curiosissimè conquirere soleam, ac libentissimè perlegere. Vix enim aliá viâ sperassem eum ad manus meas perventurum, cùm perpauca istic excusa volumi-

ipsius Ptolemæi temporibus, quibus partium .6. Martis Eccentricitatem definivit. Non possum non magnoperè probare Systema Universi á te excogitatum, quamvis cuperem Solis Orbem ac Martis nequaquam sese intersecare. Quod si, ut a studioso isto cognovi, Martem acronycum terræ propiùs accedere, quàm Solem, à te est observatum, intersecatio hujusmodi omninô admittenda est. In magna versor expectatione tuarum hujusmodi observationum & speculationum, quas & probare & sequi minimè erubescam. Etsi in construendis Ephæmeridibus resolutisque tabulis á Copernianâ ratione, & tabulis Prutenicis nè

G 4 latum

Figure 2-3 Critical passage of Magini's letter to Tycho of September 1590. It is reproduced from the elaborate folio of engravings of his famous instruments and of his "heavenly castle" that Tycho circulated to astronomers in 1598. "It's impossible to overstate how much I approve the World System you have devised," Magini writes, although he wished that it did not require the intersection of paths of Mars and the Sun. (*Courtesy of the University of Chicago Library*)

But even by 1596, both Kepler and his teacher (Maestlin) were assuring the authorities who had to approve Kepler's ardently Copernican first book that it could cause no scandal, since by now "the best astronomers were Copernican." Ptolemy continued to be taught because he was easier for beginners. These claims (in letters to the duke of Tubingen) never caused Kepler or Maestlin any trouble, although Tycho's version of a heliocentric system certainly is not what either Copernicus or Tycho would have thought of as Copernican.[16] By 1601, Kepler was writing: "Today practically no one doubts what is common to the Copernican and Tychonic hypotheses, namely that the sun is the center of the motions of the five planets."[17] Ptolemy seems to be dead. Consequently it is decidedly odd that thirty years later Galileo would frame his famous *Dialogue* as a debate on the Copernican system compared to what he always referred to as the Ptolemaic alternative. But clearly, this had something to do with the politics of the arrangement Galileo had made in a set of face-to-face meetings with the Pope before beginning. When he comes to describe the "Ptolemaic" system he has in mind, it is unmistakably Tychonic (the planets are heliocentric).[18]

Kepler's book won him a job working for Tycho. Why the ardently Copernican Kepler would go to work for Tycho has an easy explanation: Having failed to get a new university appointment, it was the best job he could find. And, indeed, it was not a bad job at all, since Kepler was certainly happy to gain access to Tycho's observations. Why Tycho would feel comfortable with an assistant who was an ardent Copernican has an odder explanation. But before the end of this chapter, that will become apparent.

On this record it is not really surprising that in 1609–1610—twenty years after Magini's support of Tycho, but still a bit before Galileo's discovery of the phases of Venus (the event commonly taken to be the crucial blow to Ptolemy)—both Kepler and Galileo published books that simply took it for granted that the significant resistance to Copernicus was Tychonic. Galileo offers only a couple of passing remarks on the world system in his report on the telescope. But his comments are specifically aimed at Tychonic opponents, ignoring Ptolemy. Kepler devotes one sentence to dismissing Ptolemy as "exploded" before going on with several pages about why Copernicus is better than Tycho.

The Geocentric Pretzels

If the success of Tycho's system did not really depend on whether or not there were solid spheres (Figure 2-2), then the neat Tychonic story has a large hole. For if the Tychonic alternative really is clearly better than Ptolemy's but there is no necessary connection with solid spheres, why did Tycho dominate Ptolemy only *after* Copernicus proposed his system? And at the other end of the usual story, how could everyone have missed the obvious (to anyone who understood Ptolemy) alternative way of providing an automatic connection without abandoning the other virtues of Ptolemy's system? But the most fundamental issue turns on what Tycho *misses* that Copernicus provides.

Figure 2-4*a* shows the paths that Ptolemaic planets must follow. That is, if you trace out where a planet goes as the machinery of Figure 1-4 or 1-5 operates, you get Figure 2-4*a*. The darker path in Figure 2-4*a* is taken directly from a diagram of the path of Mars displayed by Kepler as part of his 1609 argument for Copernicus. He invited his readers to consider whether they really could believe that the orbits of the planets were shaped like pretzels.[19]

Following Kepler's suggestion in his own caption for the pretzel path, the figure adds the paths of the rest of the planets so that you can compare the Ptolemaic (and in a moment the Tychonic) system to the paths that Copernicus exhibited (Figure 2-4*b*) in *De Revolutionibus*. The dotted circle just inside Kepler's pretzel is the Ptolemaic path of the Sun, as Kepler drew it. Outside that circle, you see the snugly nested Ptolemaic paths of Mars, which would make six loops over the twelve years, Jupiter (eleven loops), and Saturn (twenty-nine loops).[20]

If there were room inside the path of the Sun to show them, over the twelve years, you would see eight loopings of Venus, and nested inside that you would have to imagine about fifty tiny loops for Mercury.

So Kepler asks his readers whether they are really prepared to believe that this is how the world works, given that the location of the each planet on its pretzel path is always exactly where it would be if there were no actual loops at all. If the world is Copernican, what looks like pretzel-shaped paths are only parallax effects that we see because we are observing simple paths from a place that is also in orbit around the Sun.

If we consider the *inverted* Ptolemaic alternative introduced near the end of Chapter 1, nothing changes in the pretzel diagram. As discussed there, the paths will be identical, although now there will be an explanation of why the motion of each planet is exactly coordinated with the motion of the Sun. For as in Tycho's system—differing only in the scaling of the orbits—the orbits of the planets directly share a common motion with the Sun. We have a single annual motion of the whole system, not (as in standard Ptolemy) an annual motion of the Sun that in some unknown way is mimicked in the motion of each planet.

And how does Tycho improve on that? In fact, Tycho does not improve on that at all, but rather introduces complications. In a Tychonic version of Figure 2-4*a*, each planet's path would have exactly the same shape as that shown for Ptolemy. But the paths would be scaled differently, and consequently they would be located differently. For Jupiter and Saturn, the loops would be exactly the same size as those for Mars. But Jupiter's path would lie beyond an open space around the path of Mars, and then Saturn would come beyond another open space around the path of Jupiter. As viewed from the Earth, however, the size of the loops would appear to grow smaller, exactly in scale with the snugly nested loops in the Ptolemaic version of Figure 2-4*a*, as observational equivalence (Figure 1-3) requires.

But in the region of the Martian loops, a detailed picture of what actually must happen in Tycho's system would look like a bowl of spaghetti in the region of the Sun, surrounded by the decorative fringes just described for the outer planets. The circle Kepler drew for the Ptolemaic orbit of the Sun would be enlarged to cut through the loops of Mars, and sharing this same space, continually intersecting the paths of Mars and the Sun, would be the eight loops (over twelve years) of Venus and the fifty loops of Mercury.[21] Since the loopings do not return to the same point, as the diagram was elaborated to cover more years, a broad band around Tycho's orbit of the Sun would gradually fill up with ink.

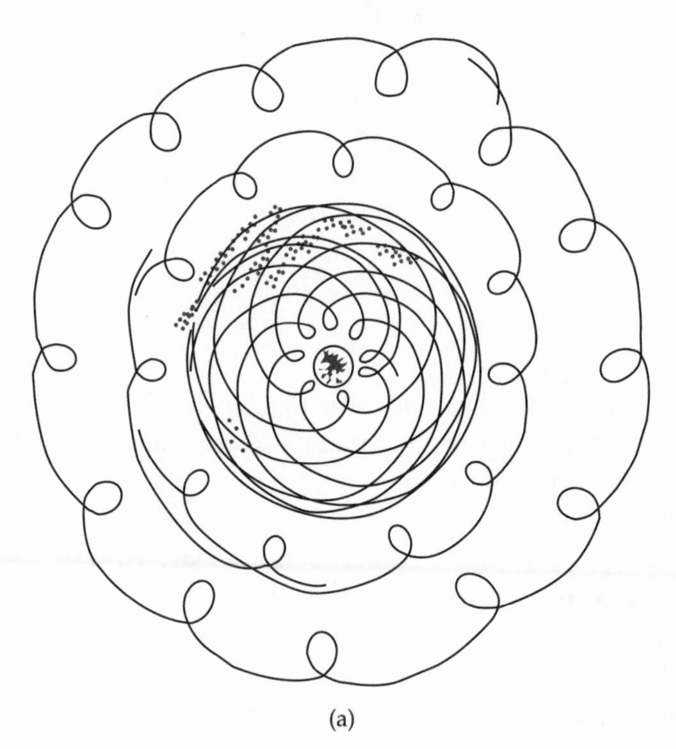

(a)

Figure 2-4 *(a)* Kepler's diagram of the geocentric path of Mars, elaborated to include Jupiter and Saturn. Looping paths for Venus and then very tight loops for Mercury would fall inside the dotted path of the Sun. In Figures 1-1 and 1-2 we see the loops from the Earth looking out; here we are looking down from high above the system, seeing the paths traced out by the Ptolemaic machinery of Figure 1-5. But from either perspective (in Figures 1-1 and 1-2 showing variation in longitude and latitude, or here showing variation in longitude and radial distance), the paths do not close and the loops continually vary in exact shape. The paths shown for Jupiter and Saturn are deliberately crude, in contrast to the exact pretzel path for Mars (from Chapter 1 of Kepler's *New Astronomy*).

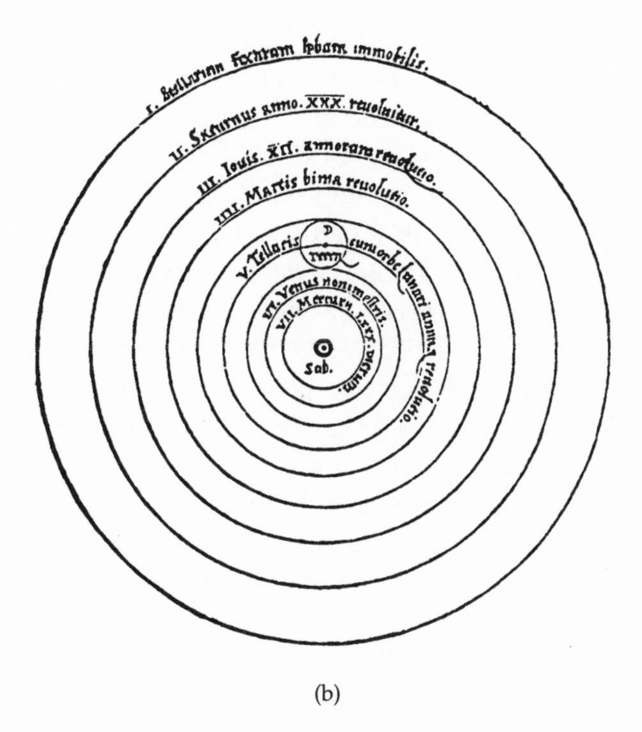

(b)

Figure 2-4 *(b)* The paths for the heliocentric world from *De Revolutionibus*. (*Courtesy of the University of Chicago Library*)

Could Tycho be sure that the Tychonic planets would not collide as they wandered through that bowl of spaghetti? Tycho raised that question himself in 1588, and correctly assured his readers that no collision was possible. And a Tychonic astronomer could indeed be sure that his planets would not collide. Why? Because Copernican planets obviously can't collide (their orbits do not intersect), and it is built into Tycho's system—indeed, effectively it *is* Tycho's system—that the Tychonic planets must always behave as if they were really Copernican!

We have a case here where a picture is indeed worth a thousand words. How easy would it be for a person to sustain the belief that the paths of the planets are exceedingly complex interwoven pretzels, but such that there will never be a collision: pretzels that can be defined only by the property that they always twist to exactly match the way Copernican orbits would appear if they were observed from a moving Earth? Kepler's comment on another aspect of Tycho's system applies conspicuously here: "Does not Nature shout out loud that the world is Copernican?"[22]

Notice that what shouts at you from these figures is not a *logical* argument for Copernicus. As a matter of logic, Kepler allowed that (see the quote in Chapter 3) God could certainly make the world in whatever way appealed to His inscrutable wisdom. But a person who understands the situation might have difficulty fending off a sense of how things work that fits the simple paths of Figure 2-4*b*, not the endlessly twisting paths of Figure 2-4*a*. Eventually pretzel paths are just not psychologically sustainable, given an alternative that produces identical observations from simple paths. Once a person starts to think of the planets moving "as if" they were Copernican, that person starts on a road that makes it ever more difficult to resist coming sooner or later to accept the idea that the world indeed might *be* Copernican.

Certainly before midcentury (and with the further push provided by what could be seen through Galileo's telescope), informed resistance to Copernicus came to be limited to sufficiently loyal Catholics and sufficiently patriotic Danes. And that is hardly surprising. This is a very stark case indeed of a disparity in economy (favoring Copernicus) in tension with a disparity in comfort (favoring Tycho). It was not easy to believe in this convoluted Tychonic miracle. But neither was it easy to believe in the Copernican notion that the Earth must be flying through the heavens. Over time, however, as both systems grew familiar, the effective contrast in economy between them became starker, and the contrast in comfort weaker.

Inverted Ptolemy

We see the first clear signs of the effect of the shift from Ptolemy to Tycho as far back as the mid-1570s. Although Copernicus's heliocentric hypothesis was at first mostly ignored, eventually there was a whole generation of astronomers who had grown up with the idea, whether they accepted it or not. In 1576 a young Englishman named Thomas Digges published a paraphrase of key passages from Copernicus, accompanied by a diagram showing a Copernican solar system set within an unbounded field of stars (Figure 2-5). The extension of the traditional compact sphere of fixed stars (as in Tycho, Figure 2-1)

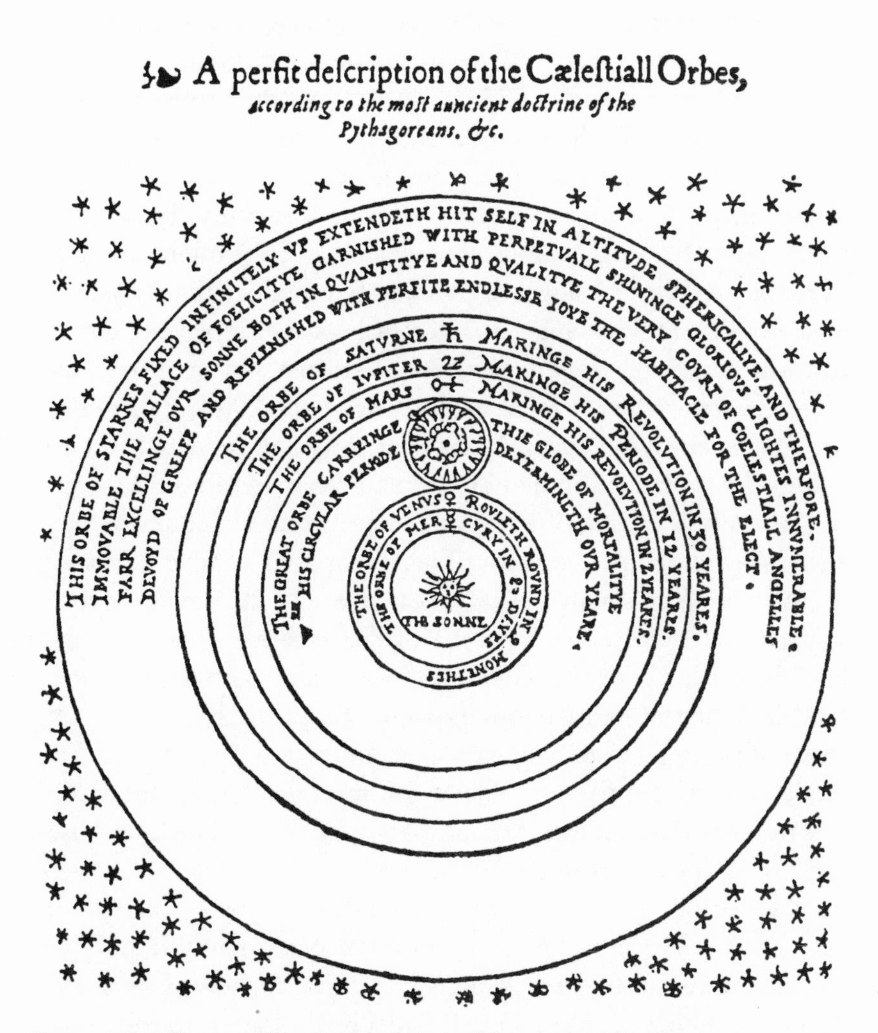

Figure 2-5 Thomas Digges's Copernican diagram (1576). Digges goes a step beyond Copernicus to explicitly show the fixed stars as spread throughout space. (*Courtesy of the University of Chicago Library*)

into the unbounded field that Digges shows makes sense for a Copernican. Gilbert (to look ahead) used almost the identical diagram, which you will see in Chapter 5. Digges was the first Copernican to step out and in bold language go beyond what was explicit in Copernicus's own discussion. It was two years after this that Tycho departed from Ptolemaic tradition and logic and made Mercury and Venus heliocentric. This was the first step toward his compromise system. But even this step destroys the snugness of the nested spheres that organizes the Ptolemaic world, for the orbits of Mercury and Venus (now both heliocentric) no longer neatly fill the presumed space between the Moon and the Sun.

So although it involved only Venus and Mercury, Tycho's early move required a psychological step away from Ptolemy that any astronomer who had not already been tempted by the Copernican idea would find wrenching. Further, making their epicycles concentric with the Sun (while leaving Mars, Jupiter, and Saturn on their Ptolemaic epicycles) gave up another element of Ptolemaic symmetry. And it also wrenched Mercury out of what appeared to be its natural geocentric position, next to the even more erratic Moon.

So only a couple of years after one young astronomer openly abandoned Ptolemy, the young Tycho proposed a move that seems to just forget the snugly nested elegance of the Ptolemaic setup described in Chapter 1. No one seems to have cared. Apparently even as early as the 1570s the Copernican idea was beginning to bite.[23]

Ten years later, astronomers were ready for the more drastic move of putting the remaining planets into heliocentric orbits, even though that entails the conspicuous intersection of the paths of Mars and the Sun that you can't miss in Figure 2-1. Tycho was intensely concerned that others might be claiming priority for the new system. In fact, several claimants were at hand. But on the record, no one defended Ptolemy against Tycho's claims. Consequently, it looks as if this new system succeeded so quickly because Tycho was seizing the driver's seat on a bandwagon that had already started to roll.

So here is the situation in 1588: The path of a Tychonic planet has exactly the same shape as the path of a Ptolemaic planet, but with the severe complication of the spaghetti bowl of intersecting paths. It also is burdened with the disruption of the most natural ordering of geocentric planets (which would put the complicated orbit of Mercury next to the even more complicated orbit of the Moon). It destroys the snug nesting of the planets by leaving unexplained gaps.[24] And most important of all: If you review the discussion of inverted Ptolemy in Chapter 1, you will see that Tycho accepts all these burdens, even though a readily available alternative also provides a linking of the Sun and the planets, and (matching Tycho's claim for his system) provides a complete ordering of the planets. It is Ptolemy's ordering, not Tycho's, but from the relevant viewpoint (that of *geocentric* astronomy) it seems plausibly right, as against a Tychonic arrangement that (from a *geocentric* perspective) looks rigorously wrong.[25]

Why, then, would astronomers around 1590, with no trace of controversy, abandon 1400 years of Ptolemaic tradition in favor of the Tychonic alternative way of producing the very same pretzel-shaped orbits? For as plausible as the Tychonic case has commonly been described, it turns out to a very strange line of argument. Even if we did not look closely at Tycho's arguments, his explanation would still fail to account for the seamless transition away from Ptolemy unless we ignore the inverted Ptolemaic option. For this simple variant on Ptolemy also provides an automatic link between the motion of the Sun and the motions of the planets. In fact, it provides the very same link.

Exactly as in the inverted Ptolemaic arrangement of Figure 1-4*b*, everything in the Tychonic world between the Moon and the fixed stars moves together. A single motion carries the orbits of the planets and the Sun. On *either* account, the Sun shares in a common annual motion with the orbits of the five planets. In Tycho's system, the planetary orbits are focused on the Sun. In the inverted Ptolemaic alternative, the planetary orbits remain in Ptolemy's nested-spheres arrangement. But the nature of the link between the Sun and the planets is *identical* either way, and except for their observationally indistinguishable difference in scaling, the paths generated by the systems are identical.

And if (though we now know that this is quite unnecessary) you think Tycho's planets follow their intricate curling paths through space without guidance from solid spheres, then inverted Ptolemaic planets (or ordinary Ptolemaic planets) could do the same. Kepler commented on how very smart Tycho's planets would have to be to manage this. But if you thought that Tycho's planets could be smart enough to do it, there was no basis for denying that Ptolemy's planets could also do it. If you thought the automatic linkage of the planets' motion to the Sun to be important (as indeed seems perfectly reasonable), that was a reason for a minor adjustment of Ptolemy (inverting the outer planets), not for abandoning the elegant Ptolemaic scheme in favor of Tycho's mess of entangled, improperly arranged orbits.

So the usual line of argument cannot actually explain the seamless move from Ptolemy to Tycho. It is as if a person living in Chicago were to drive to Milwaukee, purportedly to see a particular movie. If the movie was also playing in Chicago, then this reason for going to Milwaukee makes no sense, even if the person in fact did see the movie while in Milwaukee. We have an overwhelming reason to look for some other motive that really lies behind the trip to Milwaukee.

What came to be called the Tychonic system and what I have called inverted Ptolemy had both been available in principle all through the 1400 years that Ptolemy governed astronomy. Both can be reached from standard Ptolemy entirely with moves explicitly described by Ptolemy. In fact, the required moves were described more fully in a guide to Ptolemy (Regiomontanus, 1473), which was the mainstay of all astronomers of this period at least until Copernicus, who himself would have learned his technical astronomy from it. And in Chapter 3, we will see that Copernicus himself had put what amounts to the Tychonic system prominently in front of every reader of his book.

So by the 1580s, Tycho's system could be "discovered" by any astronomer who could read. It required no discovery at all. And other than downright dimwittedness, there is no logical reason why any astronomer who was interested in Tycho's proposal would not have noticed inverted Ptolemy as well. So why did it take 1400 years after Ptolemy, or even half a century after Copernicus, for astronomers to notice what Tycho presents as a clearly superior system when that system required no discovery? Why did no one, so far as the record shows, consider the inverted Ptolemaic alternative? And if the Tychonic arrangement was so good, why did it take so long for anyone to show an interest in it? Or, if it was not so good (as indeed it wasn't), why did his nondiscovery so quickly displace a rival with 1400 years' seniority?

The collapse of Ptolemaic astronomy at the end of the sixteenth century looks to be the scientific equivalent of the collapse of the Soviet Union at the end of the twentieth century. When long-established structures—like buildings or dams or governments—suddenly collapse, we always find that the basic framework had somehow eroded so far that the thing was ready to collapse. The same principle is at work here.

The Subliminal Appeal

Of course, to a modern reader, it just seems right that the planets should orbit the Sun. But let's ask that crucial question again: Why should it be appealing to a *geocentric* astronomer? In particular, how could Tycho's arrangement seem so overwhelmingly better than inverted Ptolemy that no whisper of discussion of the choice seems to have survived? Given the readily available inverted Ptolemaic alternative, what does this move to Tycho accomplish beyond making geocentric astronomers defend a system in which miracles might be required to avoid collisions, since the paths of the Sun, Mercury, Venus, and Mars are all tangled together?

Consider this Tychonic puzzle in terms of the tension between the two guides to intuition discussed in Chapter 1: economy and comfort. This yields an explanation that runs from B to Z. B is Otto von Bismarck, who once remarked that "of legislation and sausage, it is best not to know too much of how it is made." Z is Robert Zajonc, who in a presidential address to the American Psychological

Association argued that "emotions come first," meaning that explicit cognition is subject to powerful influence by the immediate, often unrecognizable, affective response to an object of attention.

Zajonc (1980) once arranged an experiment in which undergraduates who were distracted by overt attention to another matter were exposed subliminally to a novel tune. They were then asked to listen to a pair of tunes and indicate which they liked better. Although the subjects said that they recognized neither tune, they consistently preferred whichever tune they had subliminally been made familiar with. By and large, we like what is familiar, and certainly we feel comfortable with it. There's no place like home, we like Mom's cooking, and a thousand other examples come to mind. In the United States, easterners and westerners alike are astonished to find that people from places like Illinois actually profess to like the flat, boring landscapes of the Midwest.

From the time the heliocentric hypothesis was introduced by Copernicus in 1543, astronomers would have known that there was an alternative way to make sense of the motions of the planets. And no one who is aware of what you can see in Figures 2-4*a* and *b* can fail to notice that the Copernican system is by a wide measure the simpler way to see how the planets move. Although no one before Kepler had explicitly drawn this pretzel-shaped path, any sixteenth-century astronomer would have been at least as familiar with this contrast as Zajonc's subjects were with a tune they didn't realize they were hearing. Indeed, even many who were not astronomers would know something of these paths. Not only were the planets and stars intensely visible (with no competing electric illumination, as in the desert today), but interest in astrology, and hence in the precise location of the planets as they moved from night to night, was pervasive. Figure 2-6 shows the kind of gadgetry popular in the sixteenth century to allow a person to follow the pretzel paths as the planets varied in direction, closeness to the Earth, and (for the outer planets) brightness and apparent size.[26] Kepler would later complain that it was a shame that so much ingenuity would be wasted on such a stupid project. But to

Figure 2-6 Sixteenth-century cardboard model using nests of movable circles. The threads allowed the user to locate the center of the epicycle along the deferent (using the thread attached to the equant point), then the planet on the epicycle, then the longitude of the planet among the fixed stars. The fourth thread was used with an auxiliary device to locate the latitude. (*Courtesy of the University of Chicago Library*)

a non-Copernican, these were marvelous devices that let a person move the planets along what were taken to be their actual paths.[27]

Once the Copernican possibility was understood, it could take hold, even in the minds of astronomers who at an explicit level could not believe that the Earth was flying about the heavens. And we can in fact see that Tychonic astronomers behaved *as if* in some essential sense they had been won over to the Copernican system, even while they remained consciously opposed to Copernicanism.

For what was the appeal of Tycho's system once you notice that it has no plausible superiority relative to inverted Ptolemy? There is only one possibility. Once the inverted Ptolemaic alternative is noticed, the only clear advantage left to Tycho's system is that the Tychonic system *looks* Copernican, and the Ptolemaic system doesn't. Nothing beyond this subliminal familiarity is left to account for Tycho's success. As an explicit or even conscious reason for choice, that is idiotic. If you believe that Copernicus is wrong, what is the advantage of a contorted version of geocentric astronomy whose only real advantage is that it looks like something else that you think is radically unsound? It is not logical, but psychologically it is not hard to understand that appeal.

That no one noticed the inverted Ptolemaic alternative only replaces one puzzle with another *that has the same solution*. It would be puzzling indeed if not only Tycho but all geocentric astronomers had just by some accident failed to notice this decidedly obvious alternative.

Legislation and Sausage

So now consider the B in this B-to-Z account. To persuade someone that Tycho's system is observationally equivalent to Copernicus's, nothing more is required than simply a look at Tycho's diagram (Figure 2-1). Therefore, it is possible to follow Bismarck's advice about legislation and sausage and avoid unpleasantness by not thinking too much about how it is made. Worrying through the details of how Tychonic planets move would be more trouble than it could be worth.

With little effort, a person can see what generates the looping motions of the Ptolemaic planets. When the planet in Figure 1-5 is on the bottom of its epicycle, it is moving opposite to the main motion of the sphere carrying the epicycle. So if the epicycle is rotating faster than the sphere carrying it (both are rotating counterclockwise, but the epicycle is doing so faster), then at the bottom of its epicycle, a planet's net motion as seen from the Earth will be backward. Similarly, with the Copernican system it is easy to draw a diagram (Figure 3-2 in Chapter 3 is the diagram Galileo provided to his readers) to show how the moving Earth produces the appearance of looping motions in the other planets. But seeing how those loops arise in Tycho's arrangement is very hard. You need to envision the epicycles of the outer planets as rotating backward, that is, opposite to the direction of the annual sphere carrying the entire system. And that is not a motion we can comfortably envision.[28]

And suppose a Tychonic astronomer tried to explain the system's motions. How could that astronomer avoid being dragged into a discussion of how all the required machinery would work? That could only be awkward, which makes it understandable that the traditional commitment to solid spheres had lost its appeal. A person would have to envision a gigantic wheel carrying within it the rotating—for the outer planets, the apparently backward-rotating—set of individual orbits. Relative to the entirely incomprehensible complexity of the pretzel paths of Figure 2-4*a*, this could be a marvelous simplification, but not after Copernicus had put on the table an alternative in which that giant wheel is not needed at all. We then have only a Rube Goldberg contraption to solve a problem that just does not exist if the tiny Earth orbits the huge Sun rather than the reverse. And I have skipped over entirely the Tychonic daily rotation, 365 times as rapid, in the opposite direction from the annual motion, and on a different axis![29]

We might have supposed there would be a Tychonic tradition in which Tycho and his followers investigated puzzles related to his system with the intensity which Kepler gave to the Copernican account of planetary motion and Galileo gave the physics required in a Copernican world. But there is no such school of Tychonic work. A man like Francis Bacon could live into the 1620s and hold onto confidence in geocentric astronomy. That would not be hard for someone who was not an astronomer. Bacon, after all, was lord chancellor of England, and a lord chancellor could be well be satisfied by assurances that it was possible to account for the heavenly motions without the inconvenience of an Earth flying about the heavens. But Tychonic astronomers seemed no more interested in the details than Bacon was. Even Tycho's principle disciple (Longomontanus) paid remarkably little attention to the details of Tychonic astronomy.[30]

In fact, from start to finish Tychonic astronomy was entirely parasitic on Copernican astronomy. Tycho's system worked just like the Copernican system except that the Earth stood still. There is no puzzle at all about why Tycho would entrust his great project of making new models of the planetary motions from his elaborately exact observations to a Copernican (Kepler). There was no difference between Tychonic and Copernican astronomy, except that at the end a Tychonic astronomer pretended that the Earth stood still. Tycho certainly appeared to sincerely believe that. How many after him actually believed it has always been in doubt. After the Church forbade support of Copernicus in 1616, an adversary (Oratio Grassi) attacked Galileo for failing to support Tycho. Characteristically, Grassi did not actually defend Tycho's system. All he argued was that with Ptolemy forced aside by that "outthrust sword of Mars" and Copernicus forbidden, there was nothing else for a good Catholic to believe.[31]

A correspondent of Galileo's compared a talk Galileo gave over dinner to an insect bite: It didn't make much of an impression at the time, but when he got home it started to itch, and the itch did not go away until he took the Copernican cure. And indeed, anyone who actually thought carefully about Tycho's system could find himself turning Copernican. So we can finally make sense of the persistence over 400 years of Tycho's illusion (Figure 2-2). Modern historians never questioned it, since of course they are all Copernican, with no reason to worry over the details of how Tycho's system could work. But apparently Tychonic astronomers also did not think very hard about the system they professed to believe.

Economy, Comfort, and Tycho

So now let us return to the dual criteria of economy and comfort. As I've already mentioned, the balance between them is not static. By the mid-1570s, thirty years after the publication of *De Revolutionibus*, astronomers like Tycho, Maestlin, and Digges led a new generation that was less tightly bound to the Ptolemaic sense of how the world worked. They would have known the Copernican alternative from early in their careers, and indeed (in contrast to astronomers who were not much older), they would have been introduced to technical astronomy by teachers who were themselves young enough to have been at least a bit tempted by the Copernican alternative. The idea of an orbiting Earth would therefore not have been so uncomfortable for them as it would have been for their elders, to whom (as Copernicus himself said) it would seem a radical and indeed crazy idea.

Consider what is almost always taken to be the strongest anti-Copernican argument: the Copernican distances to the fixed stars. This has to be very large to account for the absence of observable parallax as the Earth swings around its orbit. In a *geocentric* world, it made sense to assume that the stars were embedded in a heavenly sphere. This would account for the way they maintained their never-changing relative positions as the entire field of stars moved around the Earth every twenty-four hours. And in that geocentric world, it made no sense to think that there was a huge empty space between the system of planets and the stars; on the contrary, a good part of the neatness of Ptolemy's world comes from his axiom of snugly nested spheres, with no overlaps and no wasted space. From that point of view, a field of stars that started far beyond Saturn and extended vast distances from the Earth was absurd. What could be the point of such vast empty spaces, with the nearest of stars already some large distance beyond the outermost planet?

But in a *heliocentric* world, the opposite makes perfectly reasonable sense. No fixed sphere is required. The fixed relative positions of the stars does not raise any question at all: Since in a Copernican world the fixed stars are not moving, how could they be in anything but fixed relative positions? The vast numbers of dim stars made sense if the world were huge and stars could be further suns: Dim stars could then be only very remote further suns. On an exceptionally clear night, even before the telescope, a person could see many stars that were ordinarily invisible. And that made sense in a world in which our Sun might itself be a star. But then it would also make sense (in that Copernican context) that the Sun's satellites were in its own neighborhood and far from the neighborhood of other stars.[32]

More generally, and not just in the Copernican case, we can expect to see an evolution with respect to economy and comfort for a novel proposal, and a favorable evolution if the novel idea is indeed a good one. The new approach then turns out to solve more problems than just the one that prompted its original discoverer to propose it. Thus this huge argument against Copernicus (the absence of parallax of the fixed stars) gradually faded in significance as the Copernican perspective became familiar.

Of course, things need not work out so well. Most novel proposals in fact run into messy complications, and slips in the original argument turn out to be hard to remedy. Even theories that are ultimately vindicated, like Wegener's continental drift, can go through severe trials. But a novel proposal that works gradually comes into use, and through use becomes familiar, and here familiarity breeds comfort, not contempt. And as its arguments are honed and extended, the proposal will come to seem ever more economical as well. Indeed, the novel proposals that gain followers are just those whose appeal on the economy dimension grows while their disadvantage on the comfort dimension fades. Those who try such an idea come to like it more and more.

With a wrenching choice such as Copernicus offered, where for a long time economy was sharply in conflict with comfort, many people are never persuaded. But these people die out, and after a time there are no new recruits to the old view. We are now almost ready to attend to the few people who were Copernican when it was still a bold thing to be Copernican. But having noticed the seamless and rapid transition from Ptolemy to Tycho around 1590, we should consider the plight of those who could not accept the Copernican move, but who knew the issues too well to wholly reject it.

What do people do when they are faced with a situation like that, where something that is viscerally hard to believe is also viscerally hard to doubt—where, in the language I've been using, economy is in sharp conflict with comfort? For the most part, we have to expect that they just learn to look the other way. To adapt Grassi's taunt of Galileo in a way that he would not appreciate: What else was a good Catholic to do? In prosecuting Galileo for his Copernican views, the Church—certainly the most sophisticated organization on Earth at the time, and under the leadership of a highly competent Pope—put itself in a position that remained an embarrassment for 350 years, until a Pope at the end of the twentieth century finally closed the case with a frank concession that the whole episode had been a mistake.

And what is the effect of "looking the other way" likely to be? Not good! If we compare what Tycho (or Tychonic astronomers in general) were doing in the years following 1590 with what the handful of active Copernicans were doing, the contrast is sharp indeed. Stevin, Gilbert, Kepler, and Galileo were moving toward discoveries that are celebrated to this day. Tycho, on the other hand, was adding further epicycles to his model of the motion of the Moon. It was very good technical work, yet hardly all that might be expected of a man who was convinced that he had just discovered the secret of the universe.[33]

What might a Tychonic astronomer say to all this? Certainly neither the B part nor the Z part of the B-to-Z account I've proposed is something a person would readily admit to. These are not reasons that have any appeal to reason. The wholesale abandonment of Ptolemy exhibited by the multiple claims to Tycho's system, Magini's prompt concession, the absence of any evidence of Ptolemaic resistance to Tycho, and the absence of any demand that Tycho provide the promised details of the arguments for his claims: All that makes little sense unless we assume that around 1590 there was simply a moral collapse of resistance to Copernicus.

The Copernicans' Turn

Copernicus began his book by warning his readers that he was proposing something that would seem absurd. And almost a century later, Galileo wrote of how remarkable it was that a man could so let his reason conquer his senses as to come to believe what Copernicus came to believe.[34]

It was plainly a very difficult leap to accept Earth as a planet. Anyone who managed to arrive at such a conviction would naturally become open to other radical ideas. What else that had always been taken to be so obvious that questions were simply out of order might in fact turn out to be wrong? What other astonishing things might be discovered by someone with the wit and boldness of a Copernicus? We might expect great things from such a bold questioner, especially if that questioner saw himself as part of the advance guard of a triumphant new view, rather than as an isolated, possibly deluded lone defender of what everyone else knew must be wrong. By the 1590s, though Copernicans were still vastly outnumbered, a Copernican could see that the opposition was in retreat, making a last stand in a Tychonic bastion that looked ridiculous to a Copernican. The heliocentric hypothesis was an idea whose time had come.

We might expect a good deal of exuberance from such people, and indeed that is what we see. No new empirical evidence drove Copernicus's discovery: He teased his astonishing idea out of what Ptolemaic astronomers had had in their hands for 1400 years and never noticed.[35] Once that is realized, it opens up a new world of possibilities.

Notes

1. For a summary of the usual views, see "Scientific Revolution" in the Cambridge *Reader's Guide to the History of Science* (2001), which surveys historical work on the matter and concludes with the remark that "the Scientific Revolution is no longer a historical event." Similarly, the thousand-plus pages of the *Encyclopedia of the Scientific Revolution* (2000) contain essentially nothing on the issues that are the focus of the account here.
2. The relevant group can be identified as those who were involved in the considerable surviving correspondence among the leading figures.
3. Like Galileo (and Dante), Tycho was routinely referred to in his own time, as in ours, by his first name.

4. Tycho seems to have been a rather harsh overlord, even by the standards of the time. On the other hand, he apparently was not a snob. His common-law wife came from a sufficiently modest background that no one cared to discuss it.

5. A reader with a social science background can find a nice example here of the violation of Arrow's "independence" axiom. For 1400 years before Copernicus, both little-on-big and big-on-little models of the outer planets had been available. In fact, Ptolemy had described both. But neither Ptolemy nor anyone else showed any interest in actually using the inverted (big-on-small) models, even though using them provided an automatic rather than a fortuitous link between the motion of the Sun and a motion for each planet that exactly tracks the motion of the Sun. But once a third option was available (the Copernican), those who *rejected* that option found that they now preferred the previously rejected inverted models. As with the parallel situation in social choice, what accounts for the violation of the independence axiom is that the "irrelevant" (never chosen) alternative has some appealing dimensions that, if given greater weight, would shift preferences toward that member of the original binary pair. Psychologically (not logically), the availability of the new alternative with this characteristic makes that dimension more salient, and hence can shift the choice. The salient dimension here is the importance of being able to explain rather than merely accept the presence of an annual component in the motion of each planet.

6. Tycho Brahe, trans. Hall, 1970, p. 61.

7. As discussed in Chapter 1, the ratio of the radius of the annual motion to the radius of the planet's individual motion is known directly by observation. So if there is a common radius of the annual motion of all the planets, that immediately determines the radii of the individual motions.

8. Thoren, 1990, p. 261.

9. Kuhn,1957, p. 202.

10. Tycho Brahe, trans. Hall, 1970, p. 62. There was a particular weakness in Tycho's position that was available to assist anyone who cared to defend Ptolemy, but no one took it up. Tycho claimed to have found the parallax of Mars to be larger than that of the Sun, hence demonstrating that contrary to Ptolemy, Mars comes inside the path of the Sun. The claimed observations were impossible, since the Sun is twenty times farther off than Tycho realized, and hence the correct parallax for Mars is twenty times smaller—so small, in fact, that Tycho could not

possibly have detected it. But Tycho never published his observations, which yielded his desired result only after a considerable amount of fudging. And Tycho had several years earlier assured a correspondent that the *same* observations proved that Ptolemy was right, since they showed no parallax for Mars. That no one challenged Tycho's claim cannot be explained merely by Tycho's great standing as an observer. A number of astronomers challenged his observations on other, less important, matters, as can be seen in Galileo's discussion, pp. 294–297 of his *Dialogue*, where Galileo (for once) defends Tycho. The best account of the episode is provided by Gingerich and Voelkel, 1998.

11. Tycho Brahe, trans. Hall (1970), p. 63.

12. The usual argument has been that the comet passed right through the supposed Ptolemaic spheres of Mercury and Venus, demonstrating that there are no such spheres. However, this argument is worthless as a Tychonic argument against the spheres, since in Tycho's system (although not in Ptolemy's) there is room for a comet to orbit the Sun just outside the orbit of Venus, as indeed Tycho showed it in the comet diagram that he published along with Figure 2-1. Only the tail, which was often interpreted as only an optical effect anyway, would have to penetrate another sphere. Even with respect to Ptolemaic spheres, it is easy to provide explanations that would undermine a case against the spheres. (Ptolemy himself had allowed just a bit of unexplained space between Venus and the Sun; later astronomers had adjusted this away, but of course it could have been revived. And there were other arguments available.) Consequently, Tycho's bare claim would hardly seem compelling to the writers who confidently report it, aside from the gross fact that no one today believes in such spheres. A claim that something that in fact you are sure is not there is indeed not there does not stimulate a critical concern with the details. It is uncommon to be critical of an argument whose conclusion is certainly true. But the fact that no one challenged Tycho's claim in its own time suggests that the spheres had lost their usefulness to a geocentric astronomer even then.

13. For an animation of the motions, see ftp://ftp.cogsci.soton.ac.uk/pub/psycoloquy/1998.volume.9/Pictures/munafo1.htm. But the animation should have contrasting shading for the central Earth/Moon region, as in Figure 2-2, which (like the sphere of fixed stars on the outer perimeter) would not share in the annual rotation of the single Tychonic solid sphere.

14. Magini won out over Galileo for the professorship of mathematics at Bologna, but several years later lost to Galileo for the appointment at Padua.

15. Oddly enough, Magini, in his letter endorsing Tycho's system, regrets that Tycho has found that the orbit of Mars intersects the orbit of the Sun. Even an astronomer of Magini's prominence did not seem to understand that this was completely unavoidable. However, like other astronomers toying with this idea, Tycho himself at first drew the orbit of Mars large enough to avoid the intersection, only later realizing that this would not work. That aspect of Tycho's system was clearly wrenching, which sharpens the points to come about the inverted Ptolemaic alternative, which does not involve that difficulty.

    Tycho for a time at least believed he had detected a parallax (at conjunction) for Mars of about 4.5°, which would prove that Mars came inside the radius of the Sun. But Tycho's number is 20 times larger than the actual parallax of Mars. It is what the parallax should be if the Sun's distance were what Tycho (and everyone else from Ptolemy through Tycho) thought it was. Apparently Tycho's assistant orally passed on this claim to Magini in 1590, since Magini's letter refers to it though Tycho had never published it. By printing Magini's letter, Tycho put the claim indirectly but prominently in sight, along with Magini's endorsement of Tycho's system. This strongly suggests that by that time (1598, with Magini very much alive to protest if he wished) Tycho was confident that no one would be inclined to challenge his claim. On the other hand, he did not explicitly push it either. Gingerich and Voelkel (1998) have unearthed a complicated record behind this, in which Tycho worked hard to adjust his observations to yield the parallax he wanted. He cannot have remained proud of this exercise in data fudging. Kepler apparently was not told about it, since he eventually searched for Tycho's observations supporting the claim and reported finding only a calculation of what the parallax ought to be.

16. The Kepler and Maestlin letters are printed in Kepler (1945), vol. 13, pp. 65–69.

17. From Kepler, "Defense of Tycho," in Jardine (1984) , p. 147.

18. For a detailed account, see my 1991 "Tycho's System and Galileo's *Dialogue.*"

19. The term that Kepler used was "forty days cake," which was a Lenten treat for children that eventually came to be called a pretzel (Kepler 1609). See Donahue's 1992 translation and comments.

    Kepler's diagram shows the paths generated by the Ptolemaic machinery in Figure 1-5 or (with the adjustments described in the text) the Tychonic setup in Figure 2-1. In all these figures [and in the Copernican (Figure 2-4*b*)] you see a God's-eye view, looking down on the plane of the ecliptic. Earlier, Figures 1-1 and 1-2 showed the same paths as they would actually be seen against the background of fixed stars by an observer on Earth. So they show what we see from the inside of the system looking out: and then what you see is what you see, whatever you would like to believe. Philosophers sometimes argued that the machinery of Ptolemy's (or Tycho's, or Copernicus's) models was only a mathematical fiction, however useful for "saving the phenomena." But denying the reality of Ptolemy's epicycles (or Tycho's orbits moving with the Sun, or Copernicus's moving Earth) did not alter what a person could see in the heavens: The planets continued to make their loops relative to the background of fixed stars, indifferent to what philosophers believed about the astronomers' accounts of how they did that.

20. If the planets were fixed in one place, we would have one loop per year for each, occurring as the Earth passed by in its annual motion. But since the planets are moving in the same direction as the Earth, it takes more than a year for the Earth to catch them. For Saturn, with a period of thirty years, there are consequently twenty-nine loops over thirty years, for Jupiter eleven over its period of twelve years, and for Mars one loop over each of its two-year periods.

21. Animations of the Tychonic motions of Venus and Mercury can be seen at http://www.stellium.dk/merkur2001,2.htm and http://www.stellium.dk/venus97.htm. In a Tychonic world, both performances would be going on in the same space, along with the pretzeling by Mars. At the center of the animations is the Tychonic Earth, orbited by the Tychonic Sun. Beyond what you see, the outer planets would be doing their own dances. As an amusement park thrill ride, the Tychonic motions look like vastly more fun than the boring Copernican circles. But looking at the animations can help you see why Kepler and his handful of fellow Copernicans c. 1600 could be so confident that a heliocentric world made better sense of what Copernicus called "the ballet of the planets."

The animations are from a Web site dedicated to astrology. But since the planets move as they will (relative to the Earth), this works fine whether you believe in astrology or not.

22. See Kepler's Introduction to his New Astronomy (1609).

23. Copernicus had attributed this move to Martianus Capella and other Latin writers, by way of introducing his own, vastly more radical proposal.

24. Another important disadvantage of Tycho as against standard Ptolemy, the apparently backward rotation of the orbits of the outer planets required for Tycho's scheme, is irrelevant here: This would hold also for the inverted Ptolemaic alternative way of spacing out the orbits, so it gives no net advantage either way.

25. The difference between Tycho and inverted Ptolemy is technically trivial. For inverted Ptolemy, the centers of the orbits are arranged along the Earth-Sun line on Ptolemy's nested-spheres logic. For Tycho, the identical models are also given centers on the Earth-Sun line, but they are scaled to put them all *at* the Sun. Either choice yields a definite ordering, but a different definite ordering.
26. The example shown is from Peter Apian's *Astronomicum cæsareum* (1540), which came with a set of such devices assembled and bound into the book. A handsome facsimile was published in 1967. Apian's ingenuity in constructing these devices won him appointment to the hereditary nobility, with the right (among other things) to legitimize children born out of wedlock. Cheap copies of Apian's work were made by printing the shapes and diagrams, which a user could cut out and thread together. A few much more expensive devices were produced which captured the motions in clockwork. See the plates in Poulle (1980), vol. 2.
27. Kepler's remarks come in the concluding paragraph of his Chapter 14 (p. 234 in Donahue's 1992 translation).
28. No actual backward motion is in fact required, although the intuition that backward rotation is required is not easy to overcome. And that hard-to-escape perception is uncomfortable. Things that we encounter in the world—at least things that are long-lasting—do not usually contain motions that seem to fight each other like that. If you spend a few minutes trying to see how the Tychonic system produces its required retrogressions (for the outer planets), you will get a clearer sense of how aversive it might be for a *defender* of Tycho's system to think explicitly about how that system must work.
29. Tycho (in correspondence; see Thoren 1990, p. 255) himself allowed that it might be more reasonable to attribute this daily motion to the Earth, and except where this was rigidly prohibited by obedience to Catholic doctrine, Tychonic astronomers after Tycho seem to have taken that for granted. But accepting daily rotation cured the system of a further very complicated set of motions only at the cost of vitiating all the usual Biblical and Aristotelian arguments against the Earth's motion. They all apply—indeed most directly apply—to the daily rotation.
30. See his *Astronomia Danica* (1620).
31. See Drake's 1960 collection on Galileo's controversy with Grassi. The "outthrust sword of Mars" is a reference to Tycho's report that he had measured the parallax of Mars and found that that planet thrust itself within the radius of the Sun. That cannot happen in Ptolemy's setup. Grassi's argument came in 1620, eleven years after Kepler had reported

that when he looked for these observations in Tycho's record, they did not exist. See Gingerich and Voelkel (1998) on "Tycho Brahe's Copernican campaign."

32. In replacing Ptolemy's sphere of fixed stars just beyond the orbit of Saturn with an unbounded field of stars, Digges was making a very natural conjecture for a Copernican. Copernicus himself had only alluded to it. If the Earth is in orbit around the Sun, then plainly it is the Earth that rotates east to west to provide the daily cycle, not the rest of the world that rotates west to east. Since there is no longer a need to lock the stars into a fixed sphere to explain how they maintain their exact positions, the natural conjecture is that dim stars are farther away than bright stars. And for a Copernican this is a welcome thought, since all the stars must be far beyond the outermost planet (Saturn) to account for the absence of detectable parallax.
33. There is a salient direct comparison here. Copernicus proposed not just two motions of the Earth (daily and annual), but a third motion to account for the constant orientation of its axis. Every one of our four Copernicans noticed that this third motion of the Earth was unnecessary. Therefore, in their hands, the Copernican system was shorn of an unnecessary complication. An exactly parallel idea could have shorn the Tychonic system of the backward-moving orbits it assigned the outer planets. But no Tychonic astronomer noticed that. So while all the leading Copernicans noticed how to simplify Copernicus, no Tychonic astronomer noticed how to simplify Tycho, even though the novel idea needed (a notion of inertial stability) was identical for either case.
34. *Dialogue*, p. 326.
35. Recalling the tower mentioned in Chapter 1, Copernicus was himself an observer as well as a theorist. But new observations—of which he used only a handful—only helped with updating parameters. They played no role at all in the shift from a geocentric to a heliocentric theory.

CHAPTER 3

The Discovery of Discovery

So now we reach the key question: If there was a discovery of discovery c. 1600, as Table 1-1 insists, what was it?

The most obvious element is that the revolutionary pace of discovery in science emerged in what was already an age of revolutionary discovery in technology. Printing, artillery, and clocks were creating a new world in Europe at the same time that explorers were discovering the New World across the seas. In the wake of these accomplishments of practical work came an increasing interest among bookish people in activity that got their hands dirty with something other than ink stains. Printing plays a double role here: as an example of invention and, of course, as a great facilitator of the sharing of information.

But until we come to Copernicans, c. 1600, it is hard to point to important *scientific* results from that bookish-plus-hands-on activity beyond the much admired work of Vesalius on human anatomy. And even Vesalius made no specific discovery striking enough to reasonably provide an entry for the "before 1600" side of Table 1-1. A century and more of practical discovery (and everything else that went with the Renaissance) was apparently not enough in and of itself to enable discovery in pure science. And though sometimes the claim has been pushed that activities like alchemy enabled the transition from progress in the crafts to progress in pure science, none of our four Copernicans showed much interest in such work. When they comment on it at all, their comments are routinely adverse. We have strong reason to look for what else might have been at work.

A Menu of Possibilities

Here are some possibilities:

1. Since we are looking for something that was missing until about 1600, we might start by considering apparently discoverable things that were *not* discovered until the Copernicans came along. What can we learn from *missing* discoveries?
2. Overlapping that, we might look at the sort of experiments that were done prior to the Copernicans c. 1600 to see if there was somehow a narrowness or ineffectiveness relative to what came after.
3. Attacking the puzzle from the other direction, we might consider how modern scientists work. I will sketch the usual recipe for discovery. For quite a while, philosophers tried to elaborate some version of this "scientific method" into an actual algorithm for scientific procedure. This did not succeed. Instead, it eventually produced a debunking reaction even among philosophers. But taken in a relaxed way, the recipe works! We might, therefore, scrutinize this scientific method, looking for what is in it that was somehow missing prior to about 1600.
4. And we ought to consider how far simply being a Copernican might have prompted discoveries, as a rags-to-riches story might inspire imitators, some of whom also get from rags to riches, independent of any particular following of the original's methods.
5. Finally, and central for us, we will want to consider how Copernicus himself might have come to his discovery. I've already noted that Copernicus was certainly no revolutionary with

respect to method. Here he could hardly have been a more faithful follower of Ptolemy. Yet what he discovered was logically available to any competent astronomer during all the 1400 years since Ptolemy. Since no one supposes that Copernicus was just that much smarter than everyone else, apparently circumstances somehow put him in a position to see what everyone before him had missed. On the other hand, however generously aided by serendipity, Copernicus must have *done* something different to have *seen* something so different. And perhaps his followers learned something from that.

Missing Discoveries

Turn back to near the beginning of what we can recognize as scientific discovery. The first proposal for a two-orbit model to account for the motions of the planets was developed by Appolonius in the third century B.C., as described in Chapter 1.[1] From that qualitatively promising beginning, it took 350 years before Ptolemy produced a system that actually worked. So it took as long to fill in the details of the Appolonius model as it took to go from the beginning of modern science c. 1600 to crewed flights to the Moon.

One element of that vast delay was the absence of any expectation that some manageably simple way could be found to handle the looping, never exactly repeating paths that the planets trace out against the stars. Hard work at searching requires some expectation that the search will succeed. But after Ptolemy's demonstration that the paths of the planets could be resolved into the interaction of a small set of simple motions, some expectation that parallel resolutions could turn up elsewhere could never have been entirely absent. Yet Ptolemy's example did not inspire a burst of discovery. Rather (and this is the point of Table 1-1), after Ptolemy's remarkable career, there was very little in the way of scientific discovery for 1400 years.

The lapse of time from Appolonius to Ptolemy—and then the much greater lapse from Ptolemy to Copernicus—is not so remarkable, however, once we notice how long it took to resolve various vastly more mundane questions. Right down to the time of Galileo, questions left unresolved included whether water expands or contracts when it freezes and whether a 10-pound ball of lead will fall ten times faster than a 1-pound ball, as Aristotle seems to require. No one had seriously tested Aristotle's proposal that a projectile continues its forward motion because the air pushed out of its way is in turn forced around to the back of the projectile, pushing it forward. The mathematics of parabolas had been worked out by the time of Euclid. And we might suppose that that how a projectile moves would be roughly known to any little boy who tried to see how far he could pee. But that projectiles follow parabolic paths had never been noticed.

Aristotle's own account of how projectiles move seems to have commanded little loyalty. The earliest *surviving* criticism of it dates from the sixth century, but it is easy to believe that questions were always raised about this puzzling claim. Aristotle himself did not have much confidence in it. He also proposed another possibility, although it is hard to pin down just what he meant. What is most interesting about Aristotelian physics is not that various claims were challenged but that even where (as with his account of projectile motion) a claim could be *easily* challenged, it remained a topic for disputation right down to 1600. Even a really implausible claim—if it was attributed to Aristotle—rarely seemed to be completely set aside.[2] Old questions did not die, neither did they fade away. Since the door was never firmly closed on old questions, it was also never wide open to new ones.

Early Experimenting

Unlike problems in astronomy, simple physical problems (like the puzzles about freezing water and falling objects) might seem to invite resolution by experiment. And since good experiments had in fact been done at least occasionally throughout the history of science, but simple questions were nevertheless left unresolved, we might suspect that the scope or effectiveness of experimenting was somehow limited. It is easy to find evidence of that.

Tycho lived on an island, so he had many opportunities to conduct simple experiments with how things move aboard a moving boat. Galileo taught at Padua, which was ruled by the nearby great port of Venice, so he also had ample maritime opportunities. From their reports, we might say that each saw what he wanted to see. Galileo could see wine pour straight down into a glass on a moving boat, as you have seen coffee pour straight down into a cup on an airplane that is moving vastly faster. But Tycho claimed that a cannonball dropped from the mast of a moving ship could fall into the sea behind the boat. He could hardly have observed *that* even if he had performed the experiment during a gale.

On the other hand, into the 1590s, Galileo himself believed that denser bodies fall faster than less dense ones (say, an iron ball compared to a wooden ball), and intrinsically so, not because air resistance would have a greater effect on the less dense body. He departed from Aristotle only in denying that the heavier of two *equally* dense bodies would fall faster. But, Galileo writes (around 1590), the speedier fall of a denser body is hard to observe experimentally because the less dense body falls faster at first. Hence, the denser body has to catch up and pass the lighter one before the natural difference in speed can be seen.[3]

Galileo's claim here is not only wrong but decidedly odd. It is a claim that could hardly be made without some prompting from experiments, but they can't have been very effective experiments. Eventually Galileo realized that there was something wrong with his account. He never published this blunder. But there was a considerable lapse between the time when Galileo apparently realized that this and other points in his carefully prepared early analysis of motion were wrong and his eventual discovery of the correct law of free fall.

Galileo's never-published drafts date from about 1590. A dozen years go by before we see him on his way to a solution that really works. In the interval, Gilbert's *DeMagnete* appeared. Gilbert's book was the first extended demonstration of how careful, imaginative experiments, supported by careful reasoning, could reach wholly unexpected conclusions. It was printed in 1600, and Galileo had read it by some time in 1602. Galileo later wrote that Gilbert's book had been given to him by an Aristotelian friend who was eager to get rid of it lest it contaminate his other books.[4] Galileo was by then near forty, but it is only then that he began to make the discoveries for which we remember him. We will come to that shortly. Here I only note that as late as 1590, Galileo was still capable of gross blunders about physical matters that extended to claims about experiments that could not possibly have been sustained if the experiments had been done accurately. Yet Galileo may well have been the most skilful experimenter on the planet in 1590. In wit and pugnacity, we can see in the early work that he never published that he is already the Galileo we know. But in terms of the ability to get experiments to work and the ability to make discoveries, he was no match for the Galileo who comes on stage in the early 1600s.

Others who were active in the sixteenth century were far worse. Jerome Cardano, one of the leading scholars of the time, published a demonstration of a surprising equilibrium, in which a pail of water would balance on a pole resting on a flat surface (Figure 3-1). He warned readers that the experiment would not work unless it was done with great care. For example, Cardano wrote, it is important that the pail of water have a rounded bottom. This is an experiment that had best be tried outdoors, even if the pail does have a rounded bottom. It is hard to believe that Cardano himself ever actually tried it.[5]

But by very early in the seventeenth century the situation had changed. In 1612, Galileo reported on a dispute with Aristotelian philosophers in Florence that was conducted explicitly as a debate over how to conduct and interpret experiments.[6] As we will see in Chapter 5, by now Galileo had become very effective at this. His adversaries were trying to beat him at his own game. It is not surprising that they proved no match for him. It is a surprise that so soon after the appearance of Gilbert's first extended treatment of closely analyzed experiments, the climate had changed so much that the *defenders* of Aristotelian science had implicitly accepted seeing con-

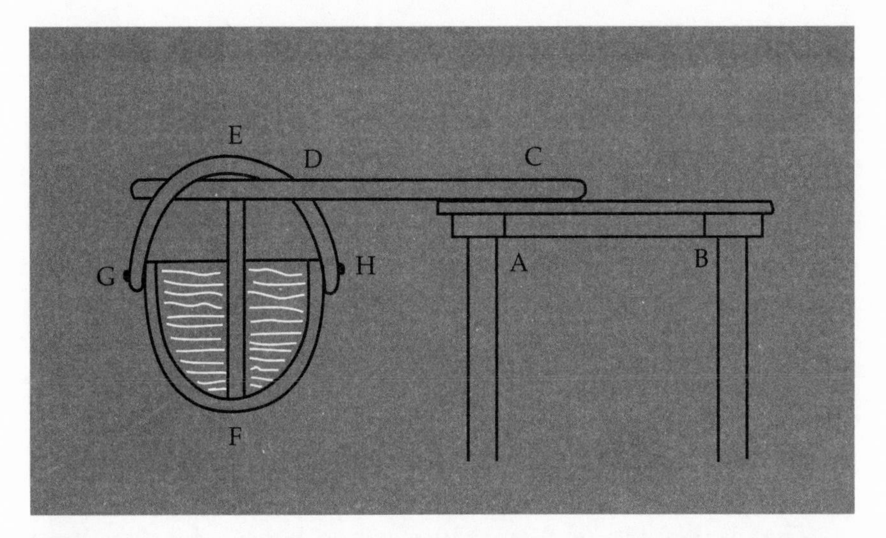

Figure 3-1 Cardano's equilibrium. He assured his readers that this would balance, provided that the experimenter was really careful. (*Courtesy of the University of Chicago Library*)

troversies in physics as being naturally settled by experiments, in contrast to an earlier tradition in which appeal to experience was common, but experience might be nothing more than whatever a person's intuition dictated. The initiative for the floating bodies contest had come from Galileo's Aristotelian adversaries, not from Galileo.

By this time all the discoveries of Table 1-1 were in hand.[7] The Aristotelians' new interest in debate with experiments as well as with words marks an important turn. But it is a response to some more fundamental turn that apparently had already occurred.

The Scientific Method

As will be seen, all the discoveries of Table 1-1 except those directly contingent on new information (Tycho's data) or a new device (the telescope) logically could have been discovered with the help of experiments feasible a thousand years or more earlier. Since experiments as part of science go back much longer, apparently something was missing from the prior *practice* of experiment. But why should it have taken 1400 years to notice that Ptolemy's models could be turned inside out to create a striking alternative system of the world? Nothing but logic is required. Apparently something was missing that not only limited the effectiveness of experimenting but in a more general way limited the potential for discovery. We might consider the sort of recipe for making discoveries that everyone today has encountered under the label of "the scientific method" as a step toward discovering what might have been missing.

But the recipe is just to be alert, look for ideas, and test the ideas. Look for intriguing aspects of what we see in the world. Mull over what can be seen until an idea appears about something that could explain what we see. And when an idea comes that looks like it might actually be made to work, pounce on it.

The idea appears as a conjecture or hunch; or, as Einstein more dramatically put it, a "free creation"; or, as Galileo put it, a possible "lucky start."[8] We commonly cannot see where the idea comes from, but somehow this step does occur. Ideas come to all of us, not just to the likes of an Einstein or Galileo. Sometimes mainly by careful analysis, sometimes by almost pure hunch, an idea will eventually appear, in the hard-to-explain way that every reader will have experienced.[9] And when the would-be discoverer turns to mulling over the idea, other ideas follow about what else ought to be true, especially what might be looked for in the world. So the original idea might then be clarified or revised or, often, abandoned. But the stage is then set for another cycle of this same recipe, but a cycle ahead of where the would-be discoverer was, one that eventually may produce something worth telling people about.

But what is in the recipe that was missing until we have Copernicans to show how to make discoveries? For unless *something* was missing, why did discoveries become common only after 1600?

Copernican Astronomy and Physics

How far might simply being a Copernican (since all four of the discoverers in Table 1-1 were Copernican) directly enhance the potential for making discoveries? Ptolemy failed to prompt anything like a scientific revolution. He turned out to be a climactic rather than a seminal figure. He concluded the long struggle to master the "problem of the planets" without inspiring any flow of further discovery. But Ptolemy's astronomy was compatible with Aristotelian physics.

On the other hand, that a Copernican world directly prompts the onset of modern science is hardly clear. A sequence of writers before Copernicus had discussed the possibility that the Earth has a daily rotation(without an annual orbit). This has an immediate intuitive appeal. Isn't it possible that it is the tiny Earth, not the rest of the world, that is rotating every 24 hours? But this far less radical possibility (relative to the annual orbit) is enough to raise nearly all the objections to a moving Earth that come up for Copernicus, and the earlier writers had already provided responses much like those that Copernicus would propose. Neither they nor Copernicus showed any real move away from the Aristotelian *style* of physics. Ptolemy argued, just as Tycho would argue, that if the Earth rotated to the east (producing only an illusion that it is the Sun rising in the east), an object dropped from a tower would fall toward the west. But Copernicus, like several earlier writers, said that if the rotation of the Earth was natural, it would naturally carry with it everything in its region, such as birds and clouds and objects falling from towers. And so on. This makes for highly unorthodox claims, but there is little reason to suppose that Aristotelian physics could not be fixed up to be compatible with Copernican astronomy, just as (as we will see in a moment) Aristotelian cosmography was quite easily fixed up to be compatible with the discovery of the New World. So Copernicanism does not plausibly directly entail the decidedly non-Aristotelian style of physics that appeared c. 1600.

However, in less direct ways, Copernican astronomy has deeper consequences. Before Copernicus, it could hardly occur to anyone to see the methods of astronomy as relevant to what went on on Earth. The two regions were made of fundamentally different sorts of stuff that followed fundamentally different natural laws. On Earth, there were four elements (earth, air, fire, and water). In the heavens, there was an entirely different element: a weightless, pure, unchanging fifth essence, called (no surprise here) the "quintessence."

Objects on the Earth moved naturally toward the center of the Earth, which was the center of the world, and only unnaturally in any other direction. Therefore, on Earth inert objects could spontaneously move only toward the center, and they would move toward this destination faster and faster until they reached some natural rate of fall or hit something that stopped that motion. Earthly objects could move in another direction only if they were violently propelled, but motion of that unnatural sort could not last once the propelling force ceased. Such violently moved objects would immediately begin to slow down and would soon stop. On this view, motion acquired by a violent push could be seen as akin to heat acquired from exposure to fire or ringing acquired when a bell is struck. All are temporary effects that begin to fade as soon as the stimulation stops.

Objects in the heavens behaved entirely differently. They did not fall, since their natural place was where they were. They did not stop moving, since eternal circular motion was what was natural for a heavenly body. All this was what everyone knew, and debate concerned only details, such as just how to understand continued movement for a while after an unnatural push had ceased.

But for a Copernican, the material of the heavens and the physics of the heavens could hardly be different from what held here on Earth, since the Earth was itself a heavenly body. Copernicus put the Earth in the heavens, which made astronomy relevant rather than irrelevant to physics on the Earth and (conspicuously for Kepler) the converse. So it was certainly more plausible in a Copernican than in a Ptolemaic world that behind the complexity directly observed on Earth, it might be possible to find exquisitely simple principles, just as the astronomers had discovered a remarkably simple account of the pretzel-shaped paths of the planets.

And encouraging such bold aspirations would be the astonishing character of what Copernicus had been led to propose. An idea that for 1800 years had seemed too obviously wrong to motivate even a discussion of why it was wrong now had begun to look convincingly right. Among the handful of people who by 1600 were impressed enough by the Copernican argument to be moved to write their own accounts of a Copernican world (that is, among our four key Copernicans c. 1600), that could be expected to have some effect, even though a number of the discoveries c. 1600 have no connection at all with Copernicanism: the hydrostatic paradox, the law of free fall, and the discoveries in optics.

But there is something else to be considered. For a Copernican c. 1600 surely would be influenced not only by what Copernicus did but also by how he managed to do it.

The Copernican Example

So now we turn finally to Copernicus himself, starting with a point that is crucial for the balance of the discussion. The revolutionary Copernican insight could not have been that the Earth might orbit the Sun, since Aristarchus had seen that 1800 years earlier.[10] Aristarchus's idea became incorporated into a well-known discussion by Archimedes. Like Zeno's paradoxes, this idea was striking enough to be remembered, whether or not it was reasonable enough to be believed. Since the heliocentric idea was never forgotten (Copernicus himself mentioned it in his draft[11]), it could not be what Copernicus discovered.

What Copernicus found was not the heliocentric idea but a reason to take that idea seriously. He saw that *if* the Earth orbited the Sun, the looping motions of the planets would be reduced to mere illusions of parallax. And yet the *logic* that in a heliocentric world the planets do not actually move in pretzel-shaped orbits requires nothing beyond what every competent astronomer had known for fourteen centuries (since Ptolemy)! The critical question then becomes: How did Copernicus see what everyone else had missed?

Remarkably little attention has ever been given to this crucial matter, probably because Copernicus himself gave it no special attention.[12] But writers are concerned with how readers will respond. Writers on highly controversial matters can be expected to avoid putting their case in a way that is likely to discomfort the readers they hope to persuade. So Copernicus's muted treatment of his elimination of the major epicycles in his otherwise very bold argument perhaps reflected the tactical problem he faced. Among his first readers, the only people who would be expert enough to follow his arguments would be those who were well versed in Ptolemaic astronomy. But that is just the sort of person in whom the nested-spheres architecture of the world described in Chapter 1 would be most firmly entrenched.

For anyone who was psychologically bound to the nested spheres, it was the major epicycles—which disappear in the Copernican system—that gave structure to the world. So Copernicus was in the awkward position of writing for readers for whom eliminating the space-filling epicycles was like kicking the legs out from under a beautifully set table. He wrote quite ecstatically of how many aspects of the planetary motions were tied elegantly together by referring the motions to his proposed orbit of the Earth. But he did not single out elimination of the epicycles for any special attention.

But of course what is relevant to the discussion here is not how people interested in astronomy were seeing things c. 1543, but how things looked c. 1600. Whatever the situation when Copernicus himself was working on the matter, over the balance of the century the Copernican sense of the heavens provided an increasingly familiar alternative way to give an order to the planets and specify their distances. It was an alternative architecture that by the end of the century was so appealing that (as we saw in Chapter 2) even geocentric astronomers could not resist it. The huge Ptolemaic epicycles were now merely the "redundancies" that Tycho announces his system will dispose of (Figure 2-1). So what Copernicus himself treated in a gingerly way, his successors c. 1600 were eager to push into the faces of their adversaries. Kepler made sure that his readers would not miss the pretzel shape of geocentric orbits (Figure 2-4*a*). Galileo points repeatedly to the centrality of this matter. Tycho claimed as a virtue of his system that he did not need these epicycles, but Kepler wants to deny Tycho that advantage by telling his readers that Tycho has not gotten rid of the epicycles, he has only hidden them.[13]

A Copernican c. 1600 would know firsthand what it takes to be persuaded by Copernicus (which we don't, since we all grew up Copernican). And the early Copernicans tell us how important it was to them that in a Copernican world, the planets do not actually travel in pretzel-shaped paths. This is certainly not the only Copernican argument. But for reasons that we have already been through in Chapter 2, this is the crucial Copernican weapon. "This alone," Galileo has Salviati remark, "ought to be enough to gain assent for the rest of the [Copernican] doctrine from anyone who is neither stubborn nor unteachable."[14] And a bit later: "It was Nicholas Copernicus who first clarified for us the reasons for this marvelous effect."[15] And a bit later yet: "I shall indeed say . . . that I have not, among the many profundities I have ever heard, met with anything that is more wonderful to my intellect or has more decisively captured my mind."[16] And from Kepler, characteristically more heart-on-sleeve than Galileo: "If God . . . had wanted the planets to execute spirals . . . he could have easily brought it about. . . . [But] what would God have preferred: that the planets should fly about in composite, ever changing, and irregular curved motions . . . or that each should describe a circle, uniform and regular as possible, clearly distinct from the confused motion that meets our eyes? . . . [T]here is no one who philosophizes soberly who would not affirm the latter opinion and altogether reject the former."[17]

Copernicus's First Step

So how did Copernicus himself come to see what Aristarchus himself and Ptolemy and everyone else had missed for 1800 years? We have advice here from an exceptionally well-qualified judge. Kepler suggests that Copernicus revealed his own path to the discovery when he advised his readers to think about putting Venus and Mercury in heliocentric orbits as a start toward putting all the traditional planets into motions around the sun. Then, he suggests, you will see how reasonable it would be to make the Earth itself a planet.

Kepler's conjecture survives in a 1601 draft *Defense of Tycho* (by this time his employer) against Tycho's hated rival, Ursus. When Tycho died, and with Ursus also dead, Kepler dropped the project. But the *Defense* survived among Kepler's papers, and in this chancy way we happen to know what Kepler thought about how Copernicus made his discovery. And in another chancy survival is independent evidence supporting Kepler's conjecture. A handwritten tabulation in Copernicus's personal copy of the Alphonsine tables shows him calculating the heliocentric distances for all the planets *except* Venus. For Venus, he just writes the number down, so apparently he had already worked that out, liked what he saw, and so was motivated to work out the numbers for the rest of the planets.[18]

The remark that Kepler points to comes at just the place in *De Revolutionibus* (I.10) where Copernicus is introducing his system of the world, so no serious reader could possibly miss it. But to see why Kepler could see moving the inner planets as the key to putting the discovery within reach, you need to try to see things from the perspective of someone who grew up in a world where knowing that the Earth is at the center of the heavenly motions was as much beyond doubt as knowing that the Sun rises in the East. Even near 1600, Copernicans appeared one at a time as converts from what they grew up knowing was right. For Copernicus the situation was far more severe, since following an argument of course is far easier than constructing the argument. To see the problem Copernicus faced, we need an effort to catch the perspective of an astronomer who was *not* already a Copernican when he made his move, but who *became* a Copernican through noticing something that no one before him had ever seen. But we might glimpse that perspective by focusing our attention on Kepler's diagram of the Ptolemaic world (Figure 1-5).

Suppose Figure 1-5 were in fact your picture of the world. And you now want to consider how things would be if, reviving that old idea from Aristarchus, you allowed the possibility that the Earth might orbit the Sun. Could you see that the retrogressions (loops) in a planet's path would become mere illusions of parallax? Please try to do that, remembering that an actual Ptolemaic astronomer would not have the advantage that you have (as a Copernican) of *knowing* that in fact the planets do not make loops.

Unless your intuition here is entirely different from mine, you will not see how the kinks disappear when you concentrate on the world of Figure 1-5. But this is the only picture a Ptolemaic astronomer could know. If you want to see that the kinks disappear, you have to get that Ptolemaic picture *out* of your head.

But if you start with Venus and move its orbit as Copernicus proposes, you come within striking distance of the great insight. Plainly Kepler thought so, and he would have been in a better position to judge than we are. And (independently, since he had no way to know of Kepler's unpublished remark) so did Galileo. In the crucial passage of his *Dialogue* where he sets up his case for the Earth's orbital motion, Galileo proceeds just as Copernicus proposed. Salviati (speaking for Galileo) starts by leading Simplicio (his Aristotelian adversary) into making Venus heliocentric, then into making the rest of the Ptolemaic planets heliocentric, and only then begins to develop an argument for making the Earth also heliocentric.[19] Simplicio agrees that the five planets at least are heliocentric, and the discussion thereafter focuses on whether that should extend to the Earth. Galileo continues to refer to the alternative as Ptolemaic, but of course having set up the choice this way, Galileo is actually considering Tycho vs. Copernicus, not Ptolemy vs. Copernicus, as the discussion in Chapter 2 would lead us to expect.

If you do what Copernicus advised his readers to do and what Galileo leads his readers to do, you can consider how things look after you make the five Ptolemaic planets heliocentric. You are challenged to think of what happens to the Earth in such a system. And since the Aristarchan proposal was familiar, you could hardly miss noticing—as Copernicus explicitly remarks—that there is a convenient empty space between the orbits of Venus and Mars into which the Earth accompanied by its orbiting moon would nicely fit. And that would put you in the position of considering what we now call the Copernican system.

Now we have a world in which there are no kinks in the planets' orbits. In this new kind of world, a person instead faces the puzzle of seeing how the kinks that are in fact observed come to be there. This is the converse of the problem faced by a geocentric astronomer, who must conceive of a structure of the heavens that generates what in a geocentric world can only be actual kinks. The Copernican system would not work unless somehow an observer would see kinks even though they are not physically there. But resolving this remaining puzzle is not hard.

From a Copernican diagram, with only some entirely straightforward further work, you can now indeed *see* how the illusion of kinks is created. Figure 3-2 is the demonstration that Galileo provided his readers. A diagram of this sort can be found in any modern astronomy text. Put ticks on a planet's orbit and on the Earth's orbit, showing how far each moves month by month, and connect the corresponding dots to see where the line they define leads against an outer circle for the background of fixed stars. The result is a diagram like Figure 3-2, where the critical insight is finally in plain sight.

But to reach that point, a person needs a sophisticated understanding of Ptolemaic astronomy. Copernicus's shift of the Ptolemaic planets into heliocentric orbits is a move from Ptolemy, not from uninterpreted observations. If you notice what Aristarchus had noticed 1800 years earlier (that the Earth might go around the Sun rather than the Sun around the Earth), but you don't have Ptolemy's space-filling nested spheres in mind, then you have nothing to weigh against the naïve intuition that the Earth doesn't do that. In Chapter 1 we noticed the parallel between the retrogressions of the planets and the apparent backward motion of a tree against a background of distant mountains as we drive past it. But if the distant mountains were merely on a painted backdrop just behind the tree, you would get no such parallax effect.

Without the help of Ptolemy's space-filling models, someone who tried to consider what it would mean if Aristarchus were right

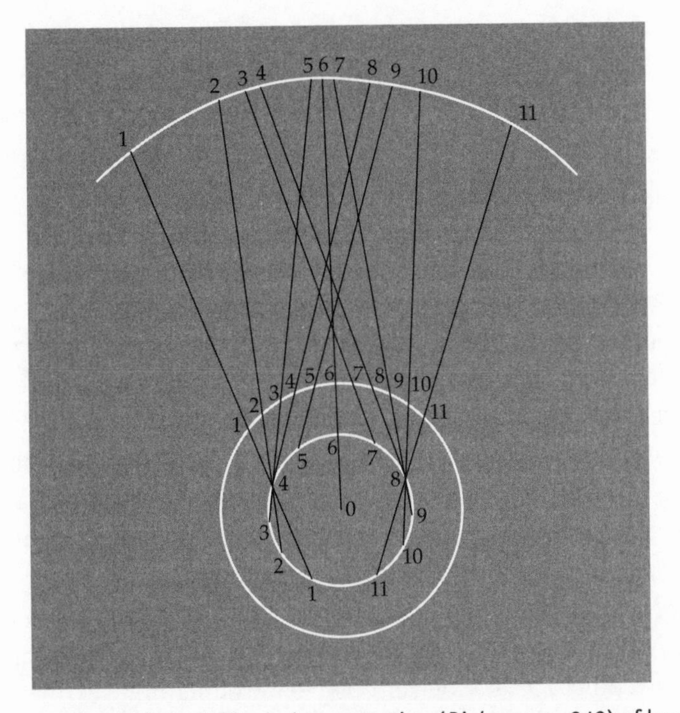

Figure 3.2 Adapted from Galileo's demonstration (*Dialogue*, p. 343) of how observations from an orbiting Earth would produce an appearance of loops in the path of another planet. A reader who looks for this should also notice (*Dialogue*, pp. 322–326) that what Galileo refers to as the Ptolemaic alternative is actually Tycho's alternative. In the geocentric alternative, as presented by Galileo, the Sun goes around the Earth, carrying the planetary orbits along with it. Mars dips inside the orbit of the Sun. *(Courtesy of the University of Chicago Library)*

would find only evidence that he was wrong. Common sense about the Earth and the absence of parallax in the stars combine to make a very convincing case. It is only with the help of Ptolemy's models that a person could discover that *if* the Earth did go around the Sun, the looping paths of the planets would be *completely* reduced to an illusion of parallax.[20] Nearly two millennia passed between Aristarchus and Copernicus, with no evidence that anyone in all that time noticed that. What it takes to see it is the combination of Ptolemy's models, Aristarchus's insight, and the long-familiar suggestion (appealing to laypeople, but not to astronomers) that Venus and Mercury might always be close to the Sun because they go around the Sun. Only when all three of these elements were combined was the great insight finally close enough to be reached.

A Window of Opportunity

So how, after 1400 years, did Copernicus, who (of course) was still a Ptolemaic astronomer when he noticed what converted him into the first Copernican, come to pull all the pieces together? How does a person come to take a deeply entrenched Ptolemaic sense of the world through two further steps, neither of which by itself produces anything memorable, but which jointly make available an unforgettable result? It is difficult to imagine what would motivate a person to take this double step, since its fruitfulness would not have been apparent until after it had been taken. So how did Copernicus think to do it?

But suppose circumstances arose in which these seemingly fruitless ideas could actually be seriously entertained for a while. From a Ptolemaic nested-spheres perspective, putting Venus and then the other traditional planets into heliocentric orbits makes no sense. As discussed in Chapter 2, that puts Mercury and Venus in the wrong order, and it leaves a hole in what had been the snugly nested Ptolemaic heavens. But if an astronomer could somehow be prompted to make that move, then thinking about how things would look from the Earth would bring the great insight into view. A person who did that might see that a conjunction of two apparently bad ideas (the heliocentric move for the planets and then the far more radical parallel move for the Earth) revealed something startling.

Somehow, apparently, circumstances led Copernicus to do all that. What is the "somehow"? Why did lightning strike just then? It turns out, remarkably, that we happen to know what could make lightening strike just then. For convincing evidence, let's date the Copernican discovery to c. 1510, thirty years before he published his book.[21] And that closely coincides with a unique window of opportunity for challenging the nested-spheres intuitions.

The window opened when Europeans first realized that what had been discovered in 1492 were not islands off the coast of China but islands off the coast of an entirely unexpected continent on the back side of the Earth.[22] In the traditional nested-spheres world view, that was simply impossible. To say that there was a second huge land mass on the far side of the world from the Eurasia/Africa land mass would be like supposing that at the back end of a horse you could find another head.

What had been taken for granted until this critical moment was an elaboration of the Aristotelian arrangement of the cosmos. At the center of the world were successive spheres of earth, water, air, and fire. Then came the heavenly spheres of Moon, the two inner planets, the Sun, and then the spheres of the three outer planets. Then came the sphere of fixed stars. And then there were the divine spheres beyond that. Moving inward, encouraged by Christian theology, the sphere of Earth was tripartite (Europe, Asia, Africa), with Jerusalem at the very center of everything. World maps printed prior to 1500 show that.

This *orbis terrarum* (the Greek *oecumene*) had to be offset a bit from the center of the sphere of water to allow a mainland on which there could be living space for human beings. God, however, could do that, even if a sufficiently rigid Aristotelian would not quite approve. In this world of nested spheres layered from Jerusalem all the way out to the highest heavens, some islands fringing the mainland could be accounted for—but a continent on the other side of the globe made no sense at all.

Columbus himself had died in 1506, still believing that he had reached the vicinity of China. But the huge outflows of fresh water from the Orinoco and later the Amazon had made it clear to sailors that what had been found had to be something vastly larger than islands off the coast of China. In 1507, a German printer named Waldseemuller published an enormous map of the world with the new continent—a second *orbis terrarum*, as Copernicus labeled it—unmistakably on view on the back of the Earth (Figure 3-3). Waldseemuller printed the map as a parcel of twelve sheets, which had to be

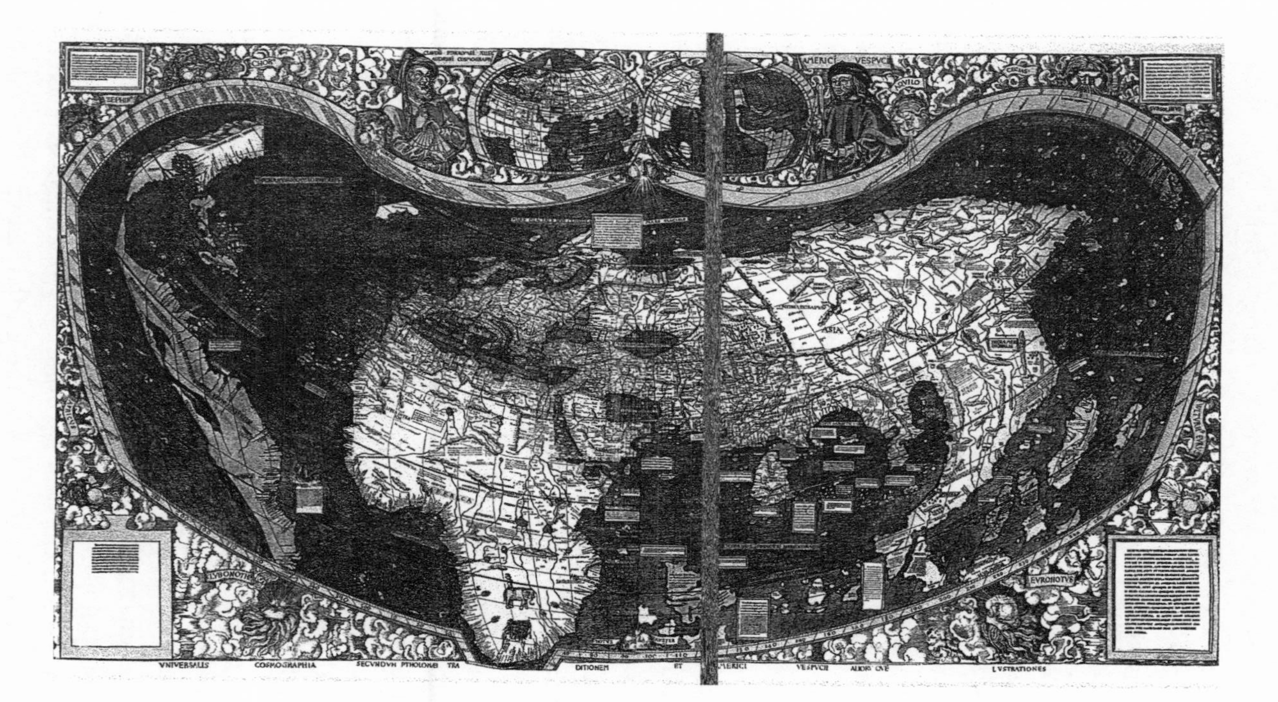

Figure 3.3 Waldseemuller's huge map (about 35 square feet) was published in twelve sheets in 1507. It was the first to clearly show a continent on the back side of the Earth, and it gave the name "America" to the new lands. Notice the inset top-center, where Ptolemy presides over the Old World, Amerigo Vespucci over the new. (*Courtesy of the University of Chicago Library*)

fitted together to see this new conception of the world. It was Waldseemuller's map that introduced the new continent to Europe and provided a name for it. In honor of what he took to be its discoverer (Amerigo Vespucci), Waldseemuller called the new continent "America." And it was just at this time, when the traditional nested-spheres sense of geography had been fatally jolted, that a Polish church administrator and amateur astronomer trying out "alternative arrangements of circles" noticed the possibility that came to be called Copernican, which abandoned the Ptolemaic arrangement of nested spheres in the heavens.

To a highly competent Ptolemaic astronomer—which is to say, to the only sort of person who could have made this discovery—the idea of moving the epicycles of Venus and Mercury to center on the Sun made no more sense than thinking that there was a continent on the back side of the Earth. But now, suddenly, the news had arrived that a new continent had in fact been found on the back side of the Earth. (Waldseemuller's map was 8 feet wide when assembled; clearly this printer thought the news was important!) For no apparent logical reason, but for a very apparent cognitive reason, Copernicus even inserted a discussion of this new continent into his book, and the language he used to describe the new discovery was taken verbatim from the captions on Waldseemuller's map, with only one significant exception. The map characterized the new discovery as an island. But Copernicus labels it "another sphere of the Earth," giving us two![23]

So here we have an indeed exceptional moment of opportunity. The stark violation of nested-spheres thinking provided by the discovery of the New World could happen only once. See Figures 3-4 through 3-7 for this window of opportunity.[24]

Even with the assistance of serendipity, why might a person consider something as bizarre as supposing that the Earth is one of the planets? What might make the heliocentric idea interesting enough for someone to suppose even for a moment that it might lead somewhere? In fact, there were three such reasons. First, the motion of the planets was certainly tied to the motion of the Sun, in the way

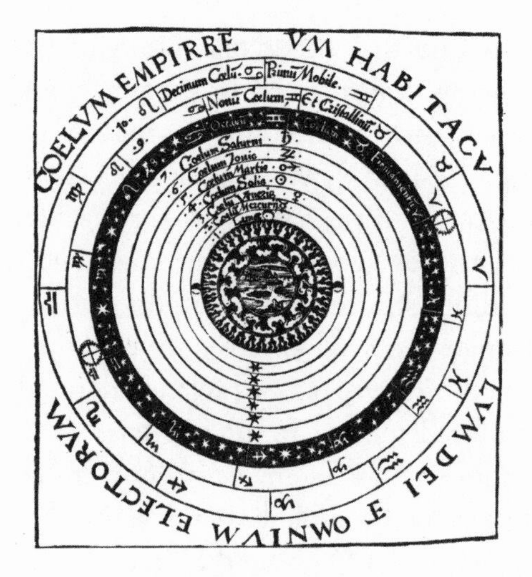

Figure 3-4 The Aristotelian nesting of a sphere of water around the sphere of Earth, but offset to allow dry land. A continent on the back side of the Earth would hardly make sense. (*From Margolis, 1987; courtesy of University of Chicago Press*)

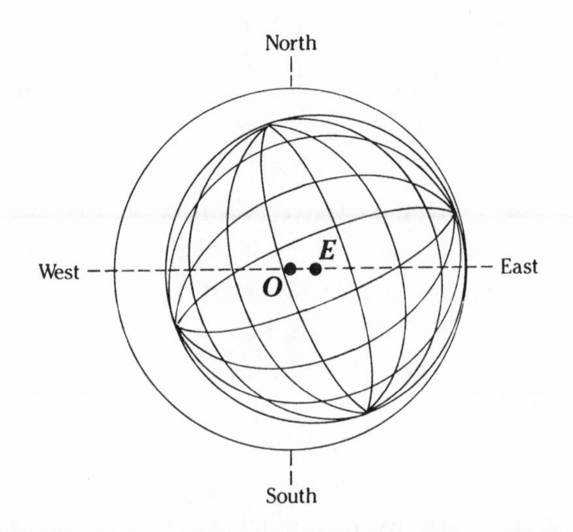

Figure 3-5 A 1493 illustration showing how this Aristotelian conception continued through spheres of air and fire into the Ptolemaic scheme of nested spheres in the heavens. Not until 1507, a year after Columbus died still believing he had reached Asia, did Waldseemuller publish the enormous map (Figure 3-3) that bluntly contradicted the traditional view of geography. (*From Hartman Shchedel's* Liber Chronicarum, *1493; courtesy of the University of Chicago Library*)

Figure 3-6 An inset from Figure 3-3 showing Amerigo Vespucci presiding over a new world that Waldseemuller decided to call "America." Logically, the Copernican discovery had been easily available for 1400 years. But actually seeing it required breaking free of the Ptolemaic nested-spheres architecture of the heavens. Indeed, it was finally seen only in the immediate aftermath of the destruction of the nested spheres in geography. This window was not open long.

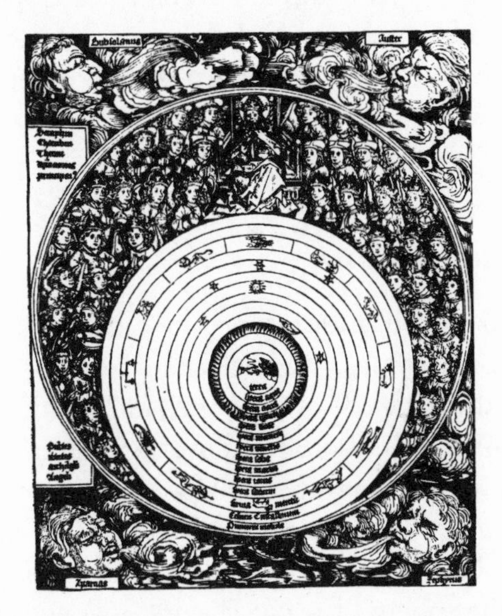

Figure 3-7 By 1523, the nesting of a sphere of Earth within a sphere of water had been quietly replaced by a single "terraqueous" sphere. But the discovery had been made, though long before the end of the century the shock to entrenched intuition that so closely coincided with it had been forgotten. (*From Peter Apian's* Cosmographia, *1523; courtesy of the University of Chicago Library*)

described in Chapter 1. As astronomers had said since long before Copernicus, "the Sun rules the planets." Second, as Aristarchus had discovered, the Sun is huge compared to the Earth, and our experience makes it easy to suspect that little things circle around big things, not the reverse. And third, as many people had noticed over the centuries, it would seem to make more sense (a much grosser case of the intuition in the second point) for the tiny Earth to turn in a daily rotation than for the whole vast world to turn daily in the opposite direction. This possibility (that the Earth, not the rest of the world, makes the daily rotation) never took hold in the face of common-sense intuitions about the stability of the Earth. But an incentive for giving some thought to the first two points would be that if it turned out that there was actually a case for supposing that the Earth has an orbital motion, you would no longer have any problem accepting also the far more immediately appealing daily rotation.

All this, however, had been available to every astronomer since Ptolemy, so these points could not account for Copernicus's seeing what everyone else had missed. What was not available to every astronomer since Ptolemy was the window of opportunity that opened with Waldseemuller's map of 1507. But looking back from 1600, how could anyone notice that window?

Looking around the Corner

The jolt to entrenched intuition provided by this second *orbis terrarum* was transient. Once it was clear that what shouldn't exist (a New World on the back side of the Earth) did in fact exist, it took less than a generation for intuitions to adjust themselves to fit what could not be denied. It took many years before people no longer found it shocking to think of the Earth as a planet, but it was vastly easier than that to replace the Aristotelian notion of a sphere of earth lopsidedly embedded in the sphere of water (Figure 3-4) with the single "terraqueous sphere" of Figure 3-7.[25] So a Copernican at the *end* of the sixteenth century, living several generations after this transition, could hardly be alert to the serendipity that seems to have enabled what

Copernicus did. The ladder having been long since removed, it looked as if Copernicus had simply scaled the wall. For by the end of the century what had once been a deep commitment to nested-spheres cosmology was so weak that we could see the uneventful abandonment of Ptolemy for Tycho that was described in Chapter 2. Nor was the evidence that the Copernican discovery coincided closely with the Waldseemuller map available until much later.

Consequently, a Copernican who was pondering how this discovery was made could only note that what Copernicus had done required somehow going beyond what could be directly seen, taking an unexpected step, looking (as I will mostly call it) "around the corner." The ladder having disappeared, looking back from the vantage point of the turn of the seventeenth century, all that could be seen was that Copernicus had had the boldness to look around a corner that no one else had thought to explore, which let him see what everyone else had missed for a very long time.

None of the early Copernicans saw looking around the corner as an explicit point of method. Yet learning how to do something very often—even most often—does *not* involve explicitly learning how it is done.[26] You probably can ride a bicycle. But if you are like a majority of bike riders, you do not know what you do to keep the bike from tipping over. But you do it. Mental habits are like physical habits. A modern reader has no occasion to marvel at the way the kinks in the paths of the planets are straightened out if the world is Copernican. Outside of a book like this, you would not even encounter the oddity that 500 years ago no one doubted that the planets follow paths with kinks in them. But if you were alive in 1600, you might find it a marvelous thing that Copernicus could turn the kinks into mere illusions, and wonder how he came to notice what everyone else for so many centuries had missed.

And mastering the argument that shows that in a heliocentric world the kinks disappear would require at least the kind of effort a student today requires to master some initially counterintuitive topic like calculus. You would have to work through the argument very carefully, and much more than once. An early Copernican, in the course of mastering the Copernican argument and then explaining and defending that argument, and also in puzzling over how Copernicus made this long-overlooked discovery, would have had occasion to think over in detail and many times how someone might come to glimpse that Copernican insight.

Historians today study figures like Darwin and Newton with great intensity, trying to understand how they came to their discoveries. Is it reasonable to suppose that our early Copernicans, actively engaged in the issue, each of them sufficiently involved to make the effort required to write a book on the matter, could have *failed* to think a great deal about how Copernicus did what he did? Such a person could absorb a sense that other fruitful things might be lying around a corner, and have the wit to realize that such a corner is hardly likely to be brightly marked "enter here."

The recipe for discovery given earlier in this chapter now can be seen to be too simple in a particular way. Finding a good idea, anyone would realize, requires effort. Aristarchus's heliocentric idea looks like a by-product of his work on estimating the size of the Sun. He was not looking for *that* idea. But the idea that he found was the by-product of work, not of idleness. The heliocentric proposal, however, did not lead anywhere until 1400 years after Ptolemy. And what seems to have been missing was comparable effort devoted to finding where to look once an idea has occurred.[27]

For a breakthrough discovery is likely to be something that somehow is at first hard to believe, as the heliocentric idea was certainly exceedingly hard to believe. Although we easily see making the planets heliocentric as almost irresistible, as the discussion in Chapter 2 showed, even for someone with no strong commitment to Ptolemy's nesting principle this is not a move that easily looks right from a *geocentric* perspective. If finding where to look around a corner were as easy as flipping a light switch, it would be hard to understand how this sort of move could have been missing until 1600. But it is not like flipping a light switch. The need for it turns discovery in science from a process that needs an epiphany (Galileo's "lucky start") into a process that might ordinarily need at least two epiphanies: we need an inspired hunch, but finding a striking idea or striking evidence or a striking new argument may require another kind of inspired hunch, this time about where to look.

Epiphanies, of course, are not easy to come by. They appear out of the blue, but not just out of the blue. They come only to people who have been persistent in thinking hard about relevant things. Hence a "where to look" epiphany is likely to reach only someone who is looking for that second sort of epiphany. But given a really spectacular example—here the most astonishing discovery that had yet been produced—a person might learn to make that effort.

Notes

1. Long before Appolonius there were schemes based on concentric spheres, with no epicycles. Various Aristotelians argued over the whole course of the following 1800 years that concentric spheres must be correct and that any use of epicycles was necessarily wrong. But a concentric-spheres model necessarily puts each heavenly body at a fixed distance from the Earth, though even casual observation shows that planets grow brighter as they approach the midpoint of their retrogressions. With a double orbit, this puzzle is automatically solved, since when the planet is at the midpoint of a retrogression, it is at the point on its epicycle closest to the Earth. And though combinations of concentric spheres can yield retrogressions, the planets follow a kind of figure-eight path rather than tracing the loops that planets actually trace out. No one was ever able to construct a concentric-spheres account that could seriously compete with Ptolemy in correctly predicting the planetary motions.
2. An instructive counterexample concerns a claim by Aristotle, but neglected by Aristotelians, that air has weight. The issue comes up in Chapter 6.
3. Drake's (1978) *Galileo at Work* provides a detailed account.
4. Galileo, *Dialogue Concerning the Two Chief World Systems*, p. 400.
5. Duhem (1991), pp. 326–329, says that Cardano seems to have misremembered a drawing of Leonardo's that showed a similar arrangement, but inverted. In Leonardo's demonstration, the center of gravity was above the table, not under it, in which case the contraption could, surprisingly, balance. As can be found in handbooks of table tricks today,

many such surprising balances, turning on the same principle, have been devised using tableware and the like. The pail of water with a rounded bottom looks like a whimsical part of such a stunt.

6. Galileo (1612) *On Floating Bodies*.
7. The main new result Galileo found in 1612 was what has come to be called the *hydrostatic paradox* (discussed in Chapter 5). This was an independent rediscovery of what Stevin had found a couple of decades earlier.
8. Galileo's phrase comes from his discussion of Gilbert (*Dialogue*, pp. 400–413). Galileo (through Salviati) says that he is going to explain Gilbert's way of doing things, which has "a certain likeness to my own" (p. 403).
9. Elsewhere (Margolis 1987, Chapter 6), I have proposed an account of how this production of unexpected insight seems to work.
10. Aristarchus found a way to calculate a distance to the Sun (actually a lower bound for such a distance), and from that and the apparent size of the Sun he was able to calculate an estimated size of the Sun relative to the Earth. Although Aristarchus's estimate was far too small, it was vast relative to anything that had previously been proposed. The Sun, he found, must be some 250 times larger than the Earth. This suggests why he could be prompted to the possibility that maybe the tiny Earth orbited the huge Sun rather than the reverse. See Van Helden 1985, Chapter 1.
11. Copernicus mentions Aristarchus in his manuscript copy. The passage that includes him was left out of the printed version, but Copernicus's characterization of the heliocentric idea as a revival of an ancient idea remained. In modern editions, the deleted material is usually included in brackets.
12. The clearest remark comes in *De Revolutionibus* 1.9, where Copernicus includes in a list of virtues of his system the point that "also, the stations, retrogressions and progressions of the planets will be seen not as their own motion but as the Earth's, which they transform into their apparent motions." Together with the other points, he says, this shows "the harmony of the whole universe, if only we look [as they say] with our eyes open."
13. See Kepler's introduction to his *Astronomia Nova* (1609), p. 51.
14. *Dialogue*, p. 342.
15. *Dialogue*, p. 345.
16. *Dialogue*, p. 356. Galileo qualifies this remark by excepting mathematical proofs, and he also puts his own exceedingly clever argument from

the apparent paths of sunspots on a par with the retrogression argument. But the sunspot argument dates only from the 1620s, and hence is irrelevant to the choices c. 1600.

17. From Kepler's draft of *Defense of Tycho against Ursus*, Jardine trans. p. 155.
18. The significance of the marginal notes was first noticed by Noel Swerdlow (Swerdlow, 1973). The interpretation here is given in more detail in "A Copernican Detective Story," Chapter 9 in my 1993 book.
19. *Dialogue*, pp. 322–326. Galileo does not treat this as a step away from Ptolemy. Galileo never describes the actual Ptolemaic system. Rather, at this point, far into the book, he is finally specifying what he has been referring to as Ptolemy's system, and it is Tychonic. I give a detailed discussion of this and other essentially political aspects of Galileo's tactics in Margolis 1991.
20. The stress on "completely" here is important. Noticing that some parallax effect would follow from an orbital motion of the Earth would hardly be convincing in the way that showing (necessarily using Ptolemy's work) that the apparent loops are *completely* an illusion of parallax would be.
21. As with most details, what we know comes to us in quirky ways. We have a piece of direct evidence, but framed in a sufficiently poetic way to leave the issue unsettled. In his preface, Copernicus, using an implicit allusion to Horace's advice to a young man to hold back his writing for nine years, says he held back his for almost four "nines." This suggests that the discovery was made soon after 1507 [1543 – (4 × 9) = 1507]. But few historians took that literally until it was discovered that a man in Krakow had an inventory of his books taken in 1514, which included what could only be a copy of Copernicus's initial sketch of his theory (the *Comentariolus*), hence dating that sketch back further than had long been supposed. On the other hand, in 1508 a friend's prefatory poem to a set of Copernican translations of a Greek author praises his work in astronomy, but in a way that seems inconsistent with his already having made the discovery. So it looks as if Copernicus got the idea no earlier than 1508, but certainly before 1514.

    The 1508 comment says that Copernicus "knew how to explain the hidden causes of phenomena on principles worthy of admiration" (trans. Dobrzyki, 1973, p. 19). But it is extremely unlikely that anyone then would have seen a claim that the Earth is flying through space as a "principle worthy of admiration." On the other hand, it would be entirely uncontroversial (then, or even a century later) to make that claim about Copernicus's use of a device (Tusi couple) that enhances the traditional commitment to uniform circular motions. And since Copernicus used that device in multiple ways (none necessarily tied to the heliocentric idea), the plural *causes* is then appropriate rather than odd, as it would be applied to the singular cause (the Earth's motion) that is the point of the heliocentric argument.

22. For a more detailed account of the "window of opportunity," see Chapters 10 and 11 in my 1987 book.

23. Rosen (1943) gives details, by way of defending Copernicus against claims that he was responsible for misnaming the New World after Vespucci instead of Columbus. The one point on which Copernicus departs from Waldseemuller's caption is that instead of calling the new continent an island, Copernicus calls it a new *orbis terrarum*, so he departs from Waldseemuller in a way that exactly fits the account here.

24. Eventually, of course, the telescope would reveal the phases of Venus, which would provide striking direct evidence that Venus was heliocentric, and hence that Ptolemy was wrong about the snugly nested spheres. This would not have come as soon as it did without the stimulus of a Copernican who was eager to see what evidence the heavens could provide. (See the discussion in Chapter 5.) But once telescopes were being built, they were bound to gradually be improved, and eventually to the point where such things could be seen. So what Galileo finally saw in 1611 was surely going to be seen by someone before the seventeenth century was out. And once the Ptolemaic structure was forced out of the picture, the heliocentric idea could be reached by direct, not around-the-corner, evidence. Of course nothing definite can be said about this hypothetical alternative history. But it is certainly not obvious that a shift to heliocentric astronomy by *this* route would have provided the kind of stimulus that occurred when (to people like Kepler and Galileo) Copernicus seemed to have just pulled this rabbit out of a hat that for 1400 years everyone else had seen as just an empty hat.

25. Randles (1980) describes the rapid shift from an off-center sphere of water almost submerging the *orbis terrarum* to a single sphere of earth with water surrounding the Earth only in its low elevations. Randles characterizes that shift as from one ancient view to another, since Ptolemy himself (in his *Geography*) describes the sea as filling hollows in a solid Earth. He finds it puzzling that neither Ptolemy nor later writers seemed to notice a contradiction between this Ptolemaic view of geography and the Aristotelian nested-spheres view. But the two perspectives do not necessarily conflict.

    Ptolemy greatly underestimated the distance of open water to be traversed by sailing west from Europe to reach Asia, a mistaken assessment that Columbus energetically embraced in seeking support for his "enterprise of the Indies." Ptolemy's world map (which can be seen in any historical book on cartography) is consistent with his "hollows" view of the oceans, but also with an Aristotelian nested-spheres view if we follow Ptolemy in supposing a narrow ocean. If you let the sphere of Earth approach the size of the surrounding sphere of water, a larger and larger *oecumene* could be accommodated, with a narrower expanse of open water on the back side of the globe, but still with no plausible possibility of a continent on the back side.

In any case, Ptolemy's *Geography* became widely available (through printing) only in 1506, so that if a Ptolemaic astronomer saw Ptolemy's "hollows" description of the sphere of water as inconsistent with his general nested-spheres account, this would have come so close in time to publication of Waldseemuller's map (1507) as to be an aspect of the "shattering of the *oecumene*."

26. There is an extensive psychological literature on tacit or implicit learning. See, in particular, Reber (1993).

27. I have put this idea first, where to look second. But it could come the other way around, which indeed (as described) is just what Kepler suggests for the Copernican moves.

CHAPTER 4

Around-the-Corner Inquiry

Opening a discussion of how science works, a prominent biologist, Richard Lewontin, remarks that "virtually the entire body of modern science is an attempt to explain phenomena that cannot be experienced directly by human beings, by reference to forces and processes that we cannot directly perceive because they are too small, like molecules, or too vast, like the entire known universe, or the result of forces that our senses cannot detect like electromagnetism, or the outcome of extremely complex interactions, like the coming into being of an individual organism from its conception as a fertilized egg."[1]

Lewontin here is making an uncontroversial point about modern science. (He tosses it off as a preliminary to his main argument.) But the kind of thing he is describing has on its face an around-the-corner character. An easy corollary of this characterization of science as about things that in their essence are out of plain sight is that discovery of such things requires persistence and cleverness in finding how to look for and how to interpret these out-of-sight effects. Persistence in working out some problem when the target *is* in sight must be primordial. It is certainly far older than the Scientific Revolution. We see it when an artisan works out a novel variant of what he already knows how to build, but also when a mathematician works out a proof of a conjecture that she already feels must be right. The persistence at issue here, however, lies in stubborn mulling and exploring about where it might be useful to look when no clear target is in sight.

In language cuter than I would wish but more evocative than anything else I have thought of, the argument here is going to be that until about 1600, inquiry was alert to direct consequences (DC), but not to consequences that might lie out of sight around some corner (AC). So before about 1600, we can find DC but not AC. Starting about 1600, we find cases of AC (as well, of course, as DC), though at first only from a few committed Copernicans.

On this account, memorable discoveries were rare for many centuries prior to 1600, because nearly everything of deep interest that was accessible to DC inquiry had already been discovered. Consequently, for a long interval after Ptolemy and Galen, there was not much left that was both memorable and available. This would account for the shortage of convincing candidates for the right-hand column of Table I-1, even though we can point to an impressive record of *practical* discoveries over the preceding centuries, and even though we can point to an impressive list of brilliant *potential* discoverers over that span of many centuries.

And as will be seen, in some way all the memorable early seventeenth-century discoveries in Table I-1 required escaping some well-established, widely shared habit of mind. So a natural conjecture is this: Suppose that what was discovered by classical science was essentially what could be noticed without violating some intuition strongly supported by experience. And suppose that indeed (the AC argument), what changed c. 1600 was that a few would-be discoverers began exhibiting a new kind of boldness and persistence in pushing intuition up against observation and analysis. What would earlier have been only transient opportunities to challenge what was taken for granted began to have some chance of turning up arguments and evidence that could overthrow even strongly entrenched intuitions. Such people, having learned to think hard about where to look, or how to contrive a place to look (an experiment), began finding what might provide the sharp challenge to set intuitions that could reassure a discoverer of being on a promising tack or persuade a skeptical audience. That could open a door to what turned out to be the vast realm of subtler around-the-corner material pointed to by Lewontin.

The "Little Boy" Case

An early reader objected to the passing remark in Chapter 3 that any little boy who tried to see how far he could pee knew something about the path of a projectile. This was ridiculous, the reader was sure. But since I did not think it was ridiculous at all, I started the account of around-the-corner inquiry in this chapter by spelling out how a former little boy alert to the potential of AC inquiry might be led to fundamental insights about how things move. This "little boy" notion was intended as merely a hypothetical illustration. But shortly after I wrote it I encountered a page from the Galileo archive (Figure 4-1) whose most straightforward interpretation would be that it sketches some version of this "ridiculous" experiment.

A former little boy thinking about puzzles of motion that had been debated since Aristotle might notice that a stream of fluid emerging from a jet (such as the stream familiar to every little boy) might follow the same path that a projectile would follow. In fact, a person alert to the possibility of evidence around the corner might notice that the stream could be thought of as a train of closely spaced projectiles. Once that oblique insight occurs to you, but of course only if it occurs to you (which is likely only if you are alert to something of this sort), it is not hard to think of how to tune up that glimmer of insight to yield measurements of the path of a projectile.

Arrange a spout so you can aim it at whatever angle you wish. Provide inflow from a tall cylinder topped off to keep the pressure constant (a technique familiar from ancient times through its use in water clocks). Now you have a stream at a measured, controllable angle emitted under a measured, controllable head of pressure. Varying the pressure gives you a set of paths for a projectile, captured as streams of water. It would then take little imagination to

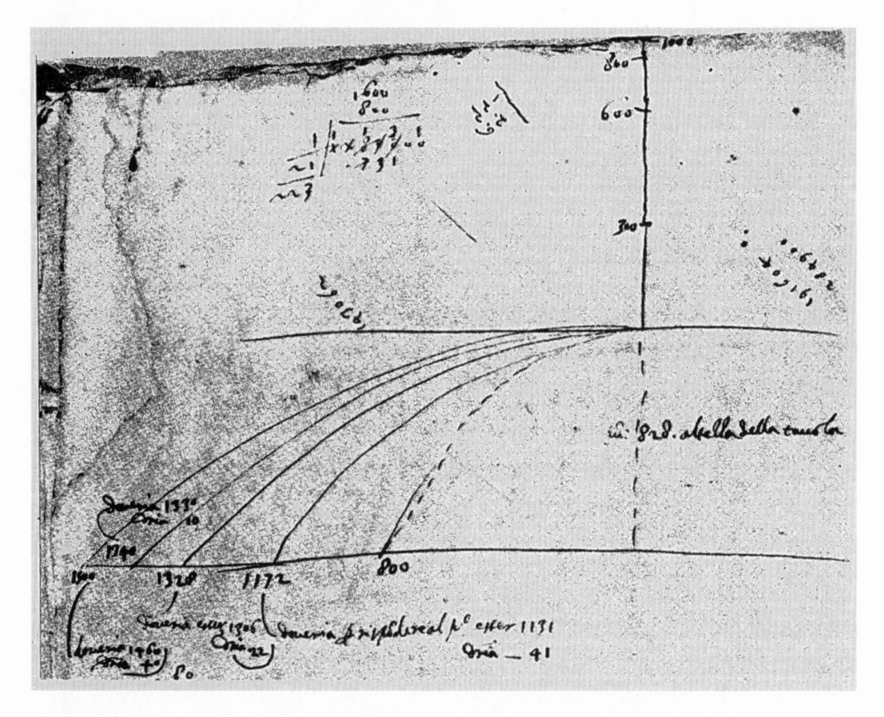

Figure 4-1 Galileo's sketch of an experiment (probably 1608) that he used in working out the parabolic path of a projectile.

think of how to record the streams on paper.[2] From there, no imagination at all would be required to notice that these paths look parabolic. And with the paths captured on paper, measurements will confirm that they are sufficiently close to ideal parabolas that it would take an unimaginative investigator indeed to fail to follow that up.

Today a parabolic curve can be described by a very simple equation, taking the horizontal motion as constant, the vertical motion as subject to a constant acceleration, and the two components as independent. For Galileo, analyzing the curve would be more complicated. But Euclidean geometry included a thorough treatment of parabolas as one of the conic sections, which was enough to put anyone who reached this "little boy" glimmer of intuition within striking distance of what we now know as Galilean inertia for horizontal motion, of the time-squared (constant acceleration) law of free fall for vertical motion, and of seeing that the horizontal and vertical velocities (in Aristotelian language, the natural and violent motions) can be treated as independent. Jointly, these three points yield the parabolic path. These interlocking results constitute the most celebrated discoveries that Galileo made in mathematical physics.[3]

But Aristotle, of course, was also a little boy once, and so was every other known student of this matter over the 2000 years between Aristotle and Galileo. And there would be numerous opportunities other than the "little boy" possibility for noticing the path of a projectile. Anyone alert for some around-the-corner way to capture that path might notice spouts from fountains, wine poured from a pitcher, waterfalls, and, under favorable conditions, fleeting but striking actual paths of single projectiles (especially flaming projectiles at night, as occurred in sieges).

So there was certainly no intrinsic lack of clues that could suggest to a person that a stream of fluid might serve as a proxy for the path of a projectile. With that insight, the rest requires only reasonable ingenuity and mechanical skill. But since no one took that kind of step for 2000 years, clearly there was something difficult about it. Somehow that kind of insight was around-the-corner—not because I have *asserted* that it was, but because if it was not, how can we make any sense at all of the 2000-year delay from Aristotle's discussion of how projectiles move to Galileo's discovery of how they move? After two millennia, just after 1600, someone finally noticed what (logically) had been readily available all along. And this long overdue breakthrough occurred when over a very short span of years a certain very small category of people (our four active Copernicans) were noticing a clump of other comparably striking things. Something happened.

Might the "little boy" approach actually be the way Galileo was led to his discoveries about motion? That it *might* be requires no verbal argument. Simply look at Galileo's sketch (Figure 4-1) of an experiment from some time before 1609. The usual interpretation is that Galileo used balls rolling down a ramp which he left out of the diagram. But Galileo's sketch can about as well (or maybe even better) be interpreted as showing streams of fluid from a column at the edge of the table.[4] Knowing for certain how Galileo obtained his measurements would not in any case tell us how he reached the insight that set him going.[5] But we can be confident of what is crucial for the argument here: that whether he started from the "little boy" insight or in some other way, Galileo indeed saw how to contrive clever means to capture the path of a projectile. And no one before had ever done that, although this topic had been discussed since Aristotle nearly 2000 years earlier.

How to Miss It

Suppose (Galileo eventually suggested) you are in the main cabin on a smoothly sailing ship. Consider what happens if you toss a ball back and forth, or let water drip from ceiling to floor, or make a standing jump toward the front then toward the rear, or watch the motion of a butterfly. Everything occurs exactly as if the ship were at anchor.[6] But anyone who was familiar with ships had experienced such things since ancient times. So why did it not become obvious to anyone who was familiar with ships that Aristotle (and everyone else who wrote about motion) must be wrong? On standard Aristotelian reasoning, an object carried by a moving ship would begin to lose its motion as soon as it ceased to be carried by the ship. By the same reasoning, an object dropped while moving horizontally would encounter interference between these contrary motions. But what you see in the cabin, and on a larger scale when an object is dropped from a mast, appears to *show* that Aristotle was wrong. Water dripping from near the ceiling to a cup on the floor does not fall astern of the cup when the ship is moving. A butterfly taking flight from the floor of its cage on a moving ship is not hit by the side of the cage, no matter how fast the ship (and hence the cage) may be moving. How could people continue to talk for 2000 years (recall Tycho and his claim about falling objects aboard ship) as if no one had had an opportunity to notice what should be obvious to anyone who has sailed aboard a ship?

But now think about why it might be easy to miss these commonplace effects. The path of a falling body is too fleeting for direct observations to clearly reveal what is happening. What you can observe from casual observation is over such short distances and times (in the cabin of a ship, for example) that perhaps there is an effect of the sort Aristotle expects, but it is too slight to be noticeable. If you wanted to really settle these issues, you would need to capture the path pretty exactly. That might seem to require such things as stopwatches and photography, which obviously were not available during that long span from Aristotle to Galileo.

That sounds reasonable. Nevertheless, it is wrong, since stopwatches and photography were not available to Galileo any more than to Aristotle. Lacking such equipment, Galileo was never able to pin down (to better than a factor of two) the *rate* of acceleration in natural fall. But even without that he was able to reach fundamental insights into the laws of motion. He saw how to contrive clever ways to control the motion of a falling object so that essential details could be caught without the help of unavailable instruments like stopwatches.

By rolling a ball down a very slightly inclined track, Galileo was able to reach the time-squared law of free fall even though he plainly lacked the means to measure the intervals of time in his formula. He realized that what was needed was not *known* intervals of time but only *equal* intervals of time. Someone who was alert to the potential of an around-the-corner move might exploit the common ability of musicians to keep in time. A careful student of Galileo's work (Stillman Drake) was able to replicate Galileo's results with the help of a briskly rendered "Onward Christian Soldiers."[7] On other occasions, Galileo measured relative times by letting water run out of a vessel and then weighing the water.

But, like the "little boy" method, beating time to music or weighing water could have been done 2000 years earlier. And although these things, like many other things Galileo did, were certainly very clever, they are not so clever that it is credible that we had to wait 2000 years for someone clever enough to see them. And about the time Galileo did this, a group of people (three other active Copernicans in a addition to Galileo) were also discovering things that, logically, some earlier clever person could have seen 2000 years before.

Cognitive Barriers

What is at issue here is what I have labeled a cognitive *barrier* (an entrenched intuition that needs to be overcome), in contrast to a logical *gap*. To an ideally rational actor, the cognitive barrier would be inconsequential unless it was complicated by a challenging logical gap.[8] This ideally rational being would not be bound by mere habits of mind. But we are not ideally rational beings. We seem to be more as well as less than that. Human beings are not good at breaking with an entrenched habit of mind even when logically we should be able to see that this habit ought not to be binding. But after a habit (whether of mind or a physical habit) is broken—if it is broken—it can be hard to recapture what it was, and it is typically impossible to recapture why it was once hard to break. After the barrier has been breached, it takes an effort to see why it ever existed—a point almost routinely illustrated in the recent literature on Copernicus.[9]

On this *barrier* argument, we should look for some intuition entrenched by familiar experience that would make it hard to see how freely falling bodies behave. In Chapter 5, you will see that Galileo himself was stopped by *some* barrier for a dozen years. But the point of the "little boy" example was to notice how utterly commonplace the clues that could provide the starting point for a really striking discovery might be. As with Galileo's shipboard examples, anyone could see it, but for a very long time no one could see what might be done with it.

Or consider a ball rolling down a slanting board. That in fact was being studied in sixteenth-century Italy. Everyday experience tells a person that the shallower the slant of the board, the more slowly the ball rolls. It requires an oblique insight, however, to see that slanting motion might reveal what happens in free fall, however easy that inference might become once the insight is in hand. It would seem to require nothing more than seeing motion on an increasingly slanted board as continuous with free fall: Eventually the board reaches vertical, and then the ball is falling free.

But as the slant is made steeper, the ball speeds up so much that its motion becomes hard to follow. Logically, the motion is continuous with motion on a minimal slant, but psychologically it seems different, and free fall seems even less continuous with the slanting motion that could comfortably be observed. Seeing balls rolling on a slanting surface as a slowed down form of free fall, or seeing a stream of fluid as a closely spaced sequence of projectiles, can *come* to seem easy. But on a 2000-year record, somehow neither was even possible prior to the arrival of our Copernicans c.1600.

How should we make sense of that enormous delay from Aristotle to Galileo? A crucial point (certainly not original here) is that Aristotle's physics was rooted in common-sense experience. Common-sense perceptions are qualitative, not quantitative. And they almost always tell a person that Aristotle was right! Heavier bodies in general do fall faster than lighter bodies. An object given a push begins losing its impetus once the push stops, just as an object that has been heated in the fire begins to cool once it is taken from the fire or a struck gong promptly begins losing its ring. If a thing is moving hard in one direction, that does interfere with pushing it away from that direction. And so on.

If you twirl a rock on a string, you cannot only see but *feel* that so long as the rock is in circular motion, it does not fall. But as soon as you stop giving it the little pushes needed to keep it twirling, it begins losing its impetus, and then it begins to respond to gravity. If you are running east to west, you cannot turn abruptly north. Somehow moving in one direction seems to diminish the effectiveness of effort to go in another direction. So two of the points essential for Galileo's claims about motion seem to violate common experience.

And in general, experience in the world makes a person feel that what Aristotle said was right. We need not suppose here that an

Aristotelian was only concerned with the surface of things, or was obligated to defend anything Aristotle said. Aristotle's physics reflect (as the first efforts to build a science surely had to reflect) how the world *feels*. What needs to be explained is not why physics started that way, but why it took 2000 years to get beyond it.

But we are all very good at neglecting things that contradict what we think must be right. The conflict between what we can see and what we believe has to be stark and persistent to defeat a solidly entrenched intuition. Galileo's shipboard effects, for example, have a fleeting or atypical character. Most of the time at sea, in fact, a ship (in Aristotle's time and long after that) either would not be moving fast enough to make the effects clear or, if it was moving fast enough, would not be sailing smoothly enough for such effects to be distinct given the rocking motions of the ship.

Apparently (since in fact no one seems to have noticed these effects until Galileo), what Nature provides to challenge Aristotle is sufficiently rare and fleeting as to be easily put out of mind unless someone is making a point of drawing attention to it, repeating it, varying it, contriving situations in which it is strikingly apparent, and in general making it no longer possible to neglect it. It is as if you were shown a clever proof that $2 + 2 = 5$. Even if you cannot see what is wrong with it, you know that it is wrong somehow and not worth taking seriously. We have a generally useful reluctance to worry about something that is of no practical importance and is in addition implausible. But thinking hard about theoretical matters that are hard to make sense of is just what AC inquiry seems to turn on.[10] In practical affairs, a sufficiently enticing target may draw us on anyway. But it looks as if until confident Copernicans were on the scene, in the realm of pure theory no one was pushed into doing that.

Francis Crick,[11] recalling how he was seized with the idea that he could discover how genes replicate, goes over the clues at hand as they were available when (with James Watson) he started to work on that. The clues that prompted Watson and Crick were not so different in effect from the heliocentric clues available to every astronomer since Ptolemy. But no one until Copernicus did anything with the clues noticed in Chapter 3. Crick's list of clues, like the astronomical clues, pointed to an intriguing conceptual question, "But the whole process seemed so mysterious one hardly knew where to begin thinking about it." Watson and Crick nevertheless began thinking about it anyway (AC inquiry), which indeed seems to have started with Copernicus.

AC Inquiry in Context

Those who took the step from direct to around-the-corner inquiry did not notice what they had done. They show no awareness of doing anything that was in principle new, although they are clearly aware that their experimenting is going further than what had been done in the past. But why should we suppose they would have been conscious of the difference stressed here between AC and DC inquiry? As I have already said, persistence in working out a puzzle and awareness that clever moves might be required is not of itself a novelty of the Scientific Revolution. But in the past there had always been a target in sight (proof of a mathematical conjecture, and so on). What is new here is persistence in looking for a "second epiphany" that could test or develop a conjecture where nothing was readily in sight. At the end of this chapter, I will sketch a Darwinian account of how that (to us) obvious extension of what was always done could be new as of 1600. But the innovators around 1600 simply saw themselves as more energetic and more careful experimenters than their predecessors and adversaries.

What is commonly at issue is isolating some particular aspect of things so that its effect can be more clearly seen. Contriving that isolation is just what is ordinarily involved in doing experiments. AC inquiry makes a very much wider range of such things available. But AC experiments, once in hand, are not qualitatively different from what was done with DC. Once you have gotten around the corner, the evidence (of course) is no longer around the corner! It is right where you can directly see it.

So AC (in the argument here) is what made the Scientific Revolution, and its results are striking and prompt, as exhibited in Table I-1; but it has the "purloined letter" character mentioned in the introduction. Before it was on the scene, no one missed it. Once it had arrived, it seemed so natural that it was hard to see. The critical work is what goes on in the head of the discoverer, and that discoverer could be aware of doing something differently only if he had access to what was going on—or not going on—in the heads of earlier investigators. Since the discoverer doesn't have this access, how could he notice that he was doing something different?

We might expect the particular discoveries of our four Copernicans to somehow grow out of particulars of their individual experience. And more than we perhaps would expect, we can at least roughly identify what those particulars were. Galileo's special interest in motion followed directly from the tradition of interest in the analysis of motion that had developed in Italy over the preceding half-century and more. That tradition had not led to anything decisive. But it put the topic in front of any ambitious young Italian investigator of nature.

Similarly, it makes sense that a member of the Elizabethan court would be seized with an interest in magnetism, a phenomenon of special salience to navigation by way of the compass. Navigation would, of course, have been a subject of paramount interest at the English court. Gilbert was the queen's physician, and when the queen was feeling well, he apparently had ample time to satisfy curiosity aroused by courtly table talk. Stevin's interest in hydrostatics seems especially appropriate for someone who as chief engineer to the prince would have some concern for the dikes in Holland. Galileo's interest in hydrostatics may by comparison then seem out of the blue. But we know that the topic was forced on him by his Aristotelian adversaries in the "debate with experiments" mentioned in Chapter 3. And Kepler's interest in optics (like that of Ptolemy and Alhazen before him) was prompted by the special concern of astronomers with understanding how to make reliable visual observations of hopelessly remote objects.

Tabletop Experiments

Challenging an entrenched intuition requires (repeating that important point) confronting that intuition with contrary experience that is too blunt to be denied. That gives experiment a special role to play. Intuitions can be challenged by mere logic, but not easily and for the most part not very successfully. As in Galileo's ship examples, ordinary experience might occasionally challenge an intuition. But if far more common or salient ordinary experience supports that intuition, then rare or transient or ambiguous contrary experiences are easily neglected. Experiment, which arranges a special experience designed to bring out clearly what is hard to make clear from casual experience, will have far more scope to be effective where AC is in play. AC inquiry can find places to look that make it hard to avoid a confrontation between what you might suppose would be true and what the world can show you if you look in out-of-the-way places. But unless you are working at finding just such things, that can hardly occur. So experiment complements AC inquiry. It is consequently not a surprise that if AC inquiry starts about 1600, so does an enlarged role for experiment.

And DC experiments turn out to be not very hard to distinguish from the AC experimenting that emerged c. 1600. DC experiments characteristically derive in some straightforward way from phenomena that occur spontaneously in "tabletop" form, where no around-the-corner move is needed. Here experimenting is only a modest elaboration of what a person might try when fiddling around with what is already right at hand.[12]

It is easiest to find early clever experiments in optics and magnetism. All that is needed is more systematic observations of phenomena that occur spontaneously: (for optics) moving glassware around to see the play of sunlight, watching a rainbow formed in the spray of a fountain, watching the refraction of a flower stem in a vase, or (for magnetism) watching what occurs when children play with a piece of lodestone. Even Galen's experiments in anatomy involved nothing beyond manipulating what was right on the table anyway, except that now the table was Galen's dissecting table.

Similarly, nothing around-the-corner is required to go from noticing where a fork balances to deliberately fiddling with a fork to see how the balance point changes if you add a bit of weight to one side or the other, or to go on to deliberately contrived experiments to find the balance points of other small objects (centers of gravity).

But it does require something oblique to see a ball rolling slowly on a tilted table as a way to examine what happens to freely falling objects. After Galileo, perhaps that was obvious, but for 2000 years before him no one had seen it. And here, in contrast to experiments that can grow directly from everyday experience with balancing a fork, or of sunlight through a glass producing a spectrum, it does not seem merely trivial to ask, "How did he think of that?"

A Taxonomy of Inquiry

In the schematic diagrams of Figure 4-2, the box on the left illustrates direct inquiry, where *A* visibly implies *B*, and *B* can be observed. The deductive direct example ($A \rightarrow B$) here could be supplemented by an inductive example $A \leftarrow B$; or *B* could be differentiated into B_1, B_2, B_3, etc. But only the basic case needs to be explicitly discussed.

In the basic case, starting from something known, *A*, plus an idea about *A*, we might be led to look for *B*. This basic case occurs when we can check what *A* plus a conjecture about *A* directly implies. If

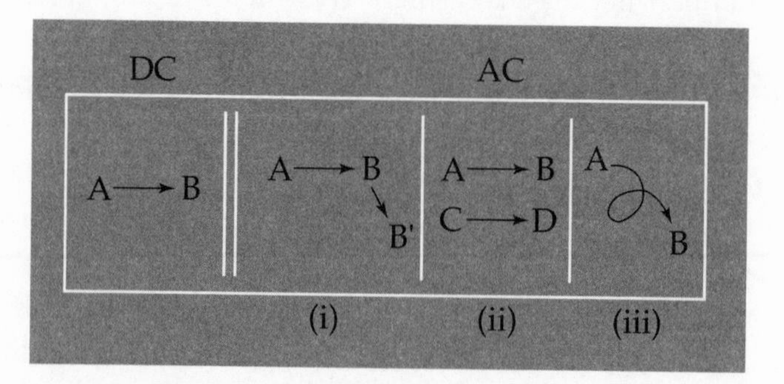

Figure 4-2 Taxonomy for around-the-corner inquiry (AC). Three variants (on the right) contrast AC with direct inquiry (DC) in the panel on the left.

that works, we are encouraged to believe that we are on the right track with the idea that led us to infer *B* from *A*; and with enough further confirmation (or failure to falsify, if you like[13]), we come to believe that our result must be right—or, if not exactly right, then at least good enough to warrant acting (for now, anyway) as if it were right. But the central point for the AC/DC distinction is that very often there is no *B* sufficiently available and unambiguous to provide a good DC test. So without AC we are left without a good test.

The around-the-corner possibilities are shown on the right of Figure 4-2. In Case i, we know *A*, and we have an idea about *A* that implies *B*. But either we are unable to observe whether *B* holds or some tests or interpretations of tests seem to show that *B* holds but others imply that it does not. Either way, we are unable to get a clear result by just directly checking *B*. But by thinking about what next might be implied by *B*, or what might jointly be implied by *A* and *B* together, we might be led to some *B′* that (unlike *B*) can be checked. *B′* was around the corner, but we found our way to where we could see it.

A second possibility (Case ii) comes if we notice C, which by analogy with $A \rightarrow B$ suggests that $C \rightarrow D$, and *D* can be checked. If we now find that on a series of points that are sufficiently striking—sometimes even one might be impressive enough—*D* indeed holds, then we have enhanced reason to take the $A \rightarrow B$ idea seriously, even though we have not yet been able to directly check *B*. In this case, we might say that the evidence was not so much around-the-corner as across-the-street. But however we label it, we have another case in which AC inquiry might uncover support for a conjecture that is not directly testable. As you will see shortly, Galileo effectively singles out this case in describing his view of Gilbert's method. He admires Gilbert's style of work, which (he says) has "a certain likeness to my own."[14]

Finally, Case iii arises when there is in fact a direct connection between *A* and *B*, but the connection can be noticed only through an argument that is convoluted or otherwise hard to see. In this case, sufficient persistence and ingenuity might bring the argument into view.

For all three cases, it is not the structure itself that identifies what is going on as AC inquiry, since, for practical problems, searching that fits each of the diagrams certainly was not new c. 1600. But in *science*, these around-the-corner searches seem to occur earlier only under fortuitous conditions, such as when (as occurs even within science) some practical problem provides an incentive for a persistent search for something that will work. The likelihood that such fortuitous conditions will occur on a matter that turns out to be important far beyond its immediate context is small—in fact (on the record), so small as to be close to nonexistent.

The Copernican discovery itself provides an example of Case i. As described in Chapter 3, we can point to a rare gift of serendipity that put Copernicus where he could get a glimpse around the corner. Looking back (from c. 1600), the connection between what Copernicus published in 1543 and the window of opportunity that opened in 1508 would be invisible. To his followers c. 1600, Copernicus had only the usual hints of a heliocentric world to work with. Looking back from 1600, it would seem that Copernicus saw more only because he looked harder. Looking directly led nowhere, since all you would see was evidence that the heliocentric idea is wrong. But exploring a side issue (just out of curiosity, how would things look if you momentarily set aside the Ptolemaic nested-spheres axiom and made the planets heliocentric?) led to a position from which a startling argument about the Earth could be seen. As we saw in Chapter 3, Copernicus did not in fact need AC inquiry to find that oblique route to his discovery, since a stunning shock to nested-spheres thinking could provoke exploring what without that assistance had always seemed to be an obviously wrong-headed possibility. But to later Copernicans who were unaware of that shock, what he did looks daringly around the corner.

And Ptolemy provided a striking example of Case ii, but without an AC move. He used measurements of refraction of an object under shallow water in a basin as the basis for approximating the refraction of starlight at low angles above the horizon. Anomalies in his observations of stars as they approached the horizon showed him something that he had to adjust for. But what looked like a parallel effect from the bending of light had long been noticed in tabletop experience with objects submerged in bowls. Ptolemy's extrapolation from tabletop refraction to atmospheric refraction was a clear case of across-the-street reasoning, but it was driven by a practical problem. It required no in-place readiness to look for such things.

And, finally, any but the simplest proofs in Euclid can provide an example of Case iii. But for ancient mathematicians as for their modern descendants, finding a proof is essentially always an exercise that has a clear target, since essentially all theorems are born as conjectures that the mathematician already feels confident enough about to motivate working at constructing a proof.[15]

So around-the-corner moves (even in science) are not invariably examples of what I am labeling AC inquiry. What seems to start around 1600, however, is searching that turns up around-the-corner moves *without* prompting either by some immediate practical difficulty or by serendipity that provokes a move that would otherwise have not been made. Hence, unlike the fortuitous cases, the search would more easily be focused on something with consequences far beyond the immediate context.

The case for a clear change would be harder to make if cases (in science) that fit the templates on the right of Figure 4-2 could easily be noticed even before 1600. With so many centuries to pick from, even if cases that happened to be focused on something that turns out to be important were really very rare, we might still find enough to blur the stark contrast in Table I-1. Yet even debatably good examples are remarkably absent. The only completely clear case is the seminal one from Copernicus himself. But examples suddenly become plentiful around 1600—all of them tied to our four Copernicans.[16]

A Darwinian Story

But how could something as fruitful as AC inquiry appear so late, and, more particularly, how could that be so when in cases of practical urgency people were capable of seeking out such moves all along? The argument here does not turn on providing a convincing answer, since that is what seems to have happened somehow whether we can see a reason for it or not. But it is tempting to sketch out a Darwinian story, and before turning to concrete examples to illustrate the taxonomy of Figure 4-2, I will do that.

Consider the conditions under which an entrenched propensity toward what I will call *second-order curiosity* could actually be a good thing. That would involve looking for where to look. And it does not seem a terribly useful propensity for a person living at the margin of subsistence in a preliterate society, which are the conditions under which our genetic propensities were shaped.

We are not likely to *inherit* a tendency that looks likely to lead to a bad end in an environment in which it would be a distraction from the hard business of survival. Curiosity would be rewarded, but curiosity about where to be curious seems pretty remote from the concerns of primitive creatures.

On the other hand, we could expect some *potential* to look for where to look. There are obviously contexts in which doing that would be unqualifiedly good for survival even if doing it would not be good in general: For example, suppose that you are trapped and your only hope is to keep thinking about how there might be some way out of this even though nothing is apparent. So a capacity for second-order curiosity (curiosity about where we might be curious) seems likely to have existed long before there were human societies that would support anything like science. But on the record, triggering second-order curiosity required some strong practical concern. That always present *potential*, however, might become much more readily accessible to someone sensitized to its usefulness, as our Copernicans might have been sensitized by the Copernican example. Then (as already described) Galileo could be moved by puzzles about motion, Kepler by puzzles about optics, and so on. And those who followed would then have examples to follow. And, less important but not trivial, in a way that never existed before, *discovery* itself could be a recognized activity, with practical rewards for someone who could make discoveries.

All this has some parallel in other human propensities (to do mathematics, play chess, write fugues, or engage in many other such stimulating exercises for the brain) that also cannot have been directly favored under hunter-gatherer conditions. With AC inquiry, as in these other domains, once such a propensity has appeared, the results could be remarkable.

A person who had somehow absorbed this alertness to clues about where exploring obscure corners might be profitable (second-order curiosity), such as the kind of person who was so taken with and familiar with the Copernican argument c. 1600 as to be moved to write a Copernican book, might be thinking hard about what else might be out there to be discovered.

Notes

1. Lewontin, 2000, p. 3.
2. For example, tilt a panel up until the stream barely grazes it, and sketch in the curve from the tracing captured. Then repeat that until the sketch matches the curve very closely. Or use a pane of glass and sketch the curve from the dry side.
3. Since any three of the points imply the fourth, I have included only three (horizontal inertia, constant vertical acceleration, and the parabolic path) in Table I-1.
4. I show only the sketch here, which is reproduced with its auxiliary notations as plate 4 in Stillman Drake's collected papers, vol. 1. As with a number of such sheets, we see only diagrams and numbers in these notes, or only bits of language to help interpret what Galileo was probably doing. On the other hand, there is no reasonable interpretation of these notes as anything but records of systematic experimenting, carefully pursued.

   I suggest (cautiously) a different interpretation from the one that Drake offers of just what Galileo may have been doing in this experiment (the controlled streams vs. rolling balls). But however it is interpreted, the diagram and its annotations are obviously a record and analysis of data from observing the paths of something put into free fall after being accelerated to varying velocities.

   The diagram shows no ramp, and the vertical scale is where it would be for the "little boy" setup and is oddly placed if what Galileo was sketching was an experiment with a ramp sloping up to the right. The scratch calculation shows Galileo relating the horizontal distance traveled during free fall to the square root of the vertical drop. In Drake's analysis, the horizontal velocity is generated by rolling a ball down a ramp not shown in the diagram. But the same calculations could also refer to the velocity of a stream under pressure from that vertical height of a column of water. Either way, the same potential energy dissipated would imply (under ideal conditions) the same relative horizontal velocities acquired. In either version, observations would be complicated by secondary effects, but the primary effect would be the conversion of potential energy at the start (some elevation above the table) into kinetic energy at the table level. The scratch calculations Galileo made in connection with the sketch seem consistent with either interpretation.

5. Galileo described (many years later, in his 1638 *Discourses*, pp. 142–143), for example, how rolling a highly polished metal ball across a highly polished, slanted metal surface leaves a visible tracing of a parabolic track.
6. *Dialogue*, pp. 186–188.
7. For a detailed account of Galileo's experimenting, see Drake's 1978 *Galileo at Work*. Drake's suggestion was that Galileo could use the analogue of frets on a guitar to elicit audible clicks on the musical beats as a ball rolled down its slightly inclined track. Then nothing more elaborate than a good ruler would be needed to measure the relation between successive intervals of time and the associated successive intervals of distance. Thomas Settle (1961), then a graduate student at Cornell, had already demonstrated that Koyre (1943) and his followers were wrong in supposing Galileo could not do the experiments he claimed by similar demonstrations for other results Galileo had reported.
8. This contrast between *barriers* and *gaps* is the key notion of my 1993 attempt to articulate what lies behind the Kuhnian insight about "paradigm shifts." A reader interested in what amounts to a cruder version of the argument here will also find that in the final chapter of my 1993 *Paradigms and Barriers*. The "ideally rational" actor here is a creature who never makes a logical error. But she has some finite processing capacity. So mistakes are never made, but there may not be time to get through all the steps needed to get from here to there; hence the possibility of a disabling logical gap.
9. The heliocentric idea is commonly treated as cognitively easy but logically dubious, as in the remark by Thoren quoted earlier. Accounts of Copernicus's advocacy of a moving Earth have often taken that to be just an incidental aspect of trying to fix the equant anomaly, or of a concern with not violating solid-spheres intuitions. But on the argument here, and (more important) on the arguments quoted from Kepler and

Galileo in Chapter 3, the opposite was true: Copernicus had a remarkably strong argument, but cognitively it was very hard to believe that the solid Earth could be flying through the heavens.

10. Greek geometry presumably grew out of physical measurements, as the etymology implies. See the discussion of the Pythagorean theorem in Chapter 6.
11. Crick, 1988, p. 35.
12. Mainly in the special case of astronomy, for someone like Ptolemy or Alhazen (each of whom had an interest in both astronomy and optics, a common overlap of interests), what is readily at hand might include instruments or techniques that gradually evolved over centuries of observation.
13. Popper's (1980) famous distinction is important, because it makes an essential, but until Popper often neglected aspect of such situations, salient. But every failure to falsify is a confirmation, and every failure to confirm falsifies.
14. *Dialogue*, p. 403. See the further discussion in Chapter 5.
15. In its time, this was as much natural science as pure mathematics. Geometry ("earth measuring") was the science of space, and theorems must have often been conjectured from experience with geometrical constructions.

    Where does the mathematician's confidence come from? To some extent always because it just "looks right," as a chess master with vast experience with the patterns that appear on a chess board will immediately have intuitions about what moves look promising enough to explore. But tacit intuition will commonly be aided by explicit analogy with other theorems already known, or by looking at what happens in transparently simple cases, or (for geometry) by what is suggested by trial constructions.
16. To fit the description here, any outstanding discovery between Ptolemy's time and the Copernican discoverers around 1600 would need to be accounted for by explanations of how it falls into the "just barely" or "serendipity" categories. I have been pretty aggressive (via posting on the Web and in talks to seminars and meetings) in looking for counterexamples, and it seems to me remarkable how little has turned up. Alhazen's theory of vision has been by far the most popular nominee. I will comment specifically on that, but the sort of comments I will make have turned out to be applicable, and usually more emphatically, to such suggestions in general.

    The c. 1600 discoveries listed in Table I-1 are striking points about the natural world that are permanent achievements. Just as all the theorems of Euclid survive to this day (although we no longer learn geometry by studying Euclid), all the discoveries in Table I-1 are fundamental things that a modern student would encounter early on, although a student would not necessarily be aware that, for example, the account of how images are formed comes from Kepler's work just after 1600, or that the "hydrostatic paradox" that underlies all hydraulic machinery was discovered by Stevin and Galileo around 1600. But no such historical slight is involved in not explicitly crediting Alhazen for his discovery, since it is not clear what he discovered (beyond what can be found in Ptolemy) that a modern student would need to learn.

This is not because Alhazen seems in any way the intellectual inferior of our Copernicans c. 1600. It is clear that he was brilliant. If we had to rank him for sheer intellect, it is hard to see why he should be put below any of our four Copernicans, even Galileo. But like Leonardo and so many other brilliant figures prior to 1600, he does not make unforgettable discoveries.

Alhazen's eleventh-century work on vision is usually (and reasonably) regarded as the most important over the long interval from Ptolemy to Kepler. Translation of his work from Arabic to Latin was very important for bringing Islamic (and indirectly Greek) science into Renaissance Europe. But if you try to specify what Alhazen discovered that was not known to Ptolemy—something that a student today would have to learn in order to understand optics—then you will have a difficult time.

Alhazen gives a particularly good analysis of the case for "intromission": the idea that we see by light coming to the eye from the object, not by the eye's shooting out rays to "see" the object (extromission). But the intromission idea can be traced all the way back to Aristotle. Alhazen certainly cannot be credited with discovering something that Aristotle had written about long before him. And what seems to us (because we know it is right) the apparent superiority of the intromission account over extromission was never *convincingly* justified until Kepler made it work. Leonardo, for example, was still partial to a role for extromission. Parallel to a key point in the Tychonic discussion of Chapter 2, what is closer to what we think today looks superior to us, even if in its own context there was nothing very convincing to commend it over its alternative. Thus Tycho's system looks to us like an advance over Ptolemy, and intromission looks better than extromission. But with respect to questions of perspective, intro- and extromission were just different ways of describing exactly the same theory. The perspective theory works equally well on either account. With respect to formation of images, a parallel point holds: Neither works, and they both fail about equally badly.

Think of an elongated pyramid with the eye at the apex and an object at the base. Straight lines from the apex to various points of the image then show the relative size of the image on a screen intermediate between the apex and the plane of the object. In some ways it is easier to think of things in terms of extromission, but since these emitted rays are invisible and apparently travel at infinite speed (as soon as you glance at the sky you can see the stars), it is easy to doubt that they exist. But on the intromission account, it is a mystery how a well-formed image finds its way into the eye. Alhazen stressed the point (making no claim to originality on this) that each point in the eye was illuminated by each point on the object. Alhazen, better than any predecessor, demonstrated this to doubters by using a specially arranged sighting device. This is one of the best-known pre-1600 experiments. But it can also serve to illustrate the contrast between AC and DC inquiry, since although the experiment is fairly elaborate, it involves no move that a person would be much tempted to call oblique or around-the-corner, given the similarity of the special device to the special devices that someone involved in astronomical observations would be accustomed to using. Nor is the idea that an illuminated point radiates in all directions counterintuitive: No one is surprised to find that aside from appearing higher in the sky, a star seen from London looks the same as that star seen from Rome.

Alhazen offered a solution (other than extromission) to the puzzle of how a precise image could be formed when the entire eye is bathed in light from each point on the object. He proposed that the eye recognized only the most *direct* ray from each point. But this is a description of what *would* work, not a convincing discovery of what *does* work, since how the eye could discriminate one ray from all the others with such precision was left as magical as how invisible extromitted rays could inform the eye of what was out there.

So for understanding *perspective*, the theories were interchangeable, and how a person preferred to describe the elongated pyramid that illustrates the perception of a distant object is just a matter of taste and convention. On the other hand, for understanding how an *image* was formed, the two views offered alternative ways of describing what still remained a considerable mystery. Both theories failed completely to adequately account for well-formed images until Kepler (just after 1600) discovered how images are actually formed: first for the case of pinhole images, as will be described, and then through lenses, including the eyeball as a lens.

I want to concede that I may well have been ungenerous to Alhazen (Ibn al-Haytham) in these comments. His work is complicated, and I do not want to pretend to be an expert on it. But reading accounts from those who are expert, I am left unclear in the way I've just described about what specific, lasting discovery should be ascribed to him.

Some other particularly impressive further examples of nicely handled experiments long before 1600 are Ptolemy's use of tabletop setups to inform his estimates of the refraction of starlight, and a discussion of magnets two centuries before Gilbert. But both can be very adequately accounted for by serendipity without AC inquiry.

Coming closer to 1600, Vesalius found many small things through his careful work at the dissection table. But here there was nothing even arguably oblique in how he made the discoveries he did make. During the fourteenth century, scholars at Oxford produced the mean speed rule for constant acceleration, which anticipated Galileo's rule for free fall by two centuries. But this work was entirely scholastic in spirit, done as a mathematical exercise with no hint that observable things in the world might be tested to see if they behaved as the rule specifies.

And the sum of what might be proposed which can survive reasonable scrutiny seems hardly likely to amount to a serious challenge to the "something happened" message of Table I-1.

CHAPTER 5

Stevin, Gilbert, Kepler, and Galileo

Of the four men who contributed to the discoveries of Table I-1, Gilbert (1544–1603) and Stevin (1548–1620) were about twenty years older than Galileo (1564–1640) and Kepler (1571–1630). And Stevin at least did his most original work about twenty years earlier, at the time that Tycho and his followers were abandoning Ptolemy for their alternative geocentric astronomy. So the four divide into two pairs of approximate contemporaries. For the older pair, we could reasonably judge that individually both Gilbert and Stevin each discovered more that has proved important for modern science than the combination of everyone who lived during the fourteen centuries between them and Ptolemy. But for Kepler and Galileo a claim this bold is not merely arguable, but beyond real dispute. If you measure what either Kepler or Galileo discovered against everything discovered in the previous 1400 years, it is no contest.

So the four divide into an early pair who are remarkable and a later pair who are even more remarkable. But the second pair has examples from the first pair to follow. None of these men ever met. But they had things in common, and knew it.

We can test whether in fact AC inquiry works essentially as I've described it by looking to see whether AC candidates in science, which are vanishingly rare before about 1600, become easy to find once we turn to the work of these four Copernicans. I will start with Stevin and Gilbert, then turn to Kepler and Galileo. The examples are all very simple, but each yielded some memorable result, and collectively they provided models for the more elaborate scientific work that followed. Illustrations of all three of the variants of around-the-corner inquiry sketched in Chapter 4 are easily found, even if we limit ourselves (as I will here) to discoveries that qualify as indisputably among the most famous of the first half of the seventeenth century.

Stevin

Stevin's parents were wealthy, but unmarried, which suggests that his upbringing may have been odd. He seems to have started out as a bookkeeper, a more prestigious profession then than it is now. But he branched out into everything else. He eventually was tutor to the crown prince, then chief engineer when that prince came to power. The *Dictionary of Scientific Biography* describes Stevin's books as covering "mathematics, mechanics, astronomy, navigation, military science, engineering, music theory, civics, dialectics, bookkeeping [of course], geography, and house-building."

Stevin provided the first important new results in hydrostatics since Archimedes, which he published in 1586. The book was not available to readers who did not know Dutch until 1605, although the startling demonstration that he added as an appendix certainly might have been known across the Channel even if the book was not. At least some of his work was well known, and no doubt more would have been had Stevin not been convinced that there was no language like Dutch for science. He wrote a pamphlet on that as well (which of course was not translated). His pamphlet on the advantages of decimal fractions popularized that idea in Europe, although it would be many years before his extension of the idea to currency and units of measurement would win out.

Stevin was especially pleased with his at-a-glance demonstration of forces on inclined planes, which he used as a colophon for his books and which eventually found a place on his gravestone (Figure 5-1). Merely examining this elegant arrangement convinces a viewer (without any need to actually do the experiment) that the effective weight of the two balls resting on the right side of the triangle must just balance the effective weight of the four balls on the left. Otherwise, in the absence of friction, the chain would go around in perpetual motion, and every viewer has an immediate conviction that that could not possibly be right. No one who looks at the diagram supposes that perpetual motion would ensue if the arms could be made smooth enough.

The AC move here is not replacing single weights on either side of a lever with a chain of balls. Stevin here was following what

Figure 5-1 Stevin's at-a-glance demonstration of equilibrium on an inclined plane. The caption says "wonder that it is no wonder," playing on the ambiguity of *wonder* as either "amazing" or "bewildering." We might say, "How amazingly simple!"

Archimedes had done in the course of providing his own proof of the law of the lever 1800 years earlier. The AC move is going from considering equilibrium between weights on a wedge to considering a necklace draped over the wedge. Prior accounts, going back to Archimedes, of what today we call the mechanical advantage of an inclined plane turned on much more complicated arguments. Stevin makes the around-the-corner move of demonstrating the mechanical advantage of an inclined plane by just looking at what you see if you drape a necklace over a wedge.

Notice that what makes this mere thought experiment so immediately convincing is the complete loop of the chain, although only the weights that are directly on the wedge are actually at issue. If all Stevin had shown had been a length of chain sufficient to cover the two arms of the incline, it would not be immediately obvious that the chain is in equilibrium. Might it not instead slide off to the side with the longer length? It is the *irrelevant* loop hanging below the block, which in itself is obviously in equilibrium, that makes Stevin's point compelling. For now you can see that unless the relevant portions (those lying on the arms of the wedge) were in equilibrium, perpetual motion could ensue, which you know cannot be right. It is then a simple matter to notice that the downward force of a heavy body on a slanting surface must in general be in proportion to the sine of the slant angle.[1] An around-the-corner move, yielding a very direct result.

In this case, there is no discovery—there is just an around-the-corner demonstration of something that was already known. But like so many of the discoveries that appear in the burst around 1600, this elegant demonstration calls on nothing that had not been perfectly familiar for 1800 years. Nothing was ever lacking but the ability to see what was logically right at hand, but which required an AC move to notice.

The Hydrostatic Paradox

The most famous discovery of the most admired mathematician of ancient times was Archimedes's demonstration that a body in water loses an amount of weight just equal to the weight of the volume of water displaced. Although this was indeed a celebrated discovery, logically it was hardly a difficult discovery. After all, anyone can see that a boat rides higher in the water when empty than when loaded: so clearly a given floatable volume displaces more water when heavier. And anyone who has lifted an anchor knows that things weigh less under water. So (logically) anyone who was interested in how far a body will sink into the water could have discovered what Archimedes discovered by very simple and direct experiments. All it takes is common experience, though on the record it also helps to be Archimedes.

Start with a tub (or just a cup) that is filled to overflowing. Add a floatable object. A volume of water that obviously must just equal the volume of whatever fraction of the solid object has sunk into the water will overflow. Then comparing the weight of the water displaced with the weight in water of the object shows that the weight that an immersed body loses is just the weight of the water it displaces. If a body is less dense than water, then before it has sunk completely into the water, it will have already displaced its entire weight, so of course it now floats. Why should it sink further if it now weighs nothing? And a body that is denser than water also weighs less in the water than it does out of the water by just the weight of the volume of water it displaces; however, since the object still has some weight, it goes to the bottom.[2] Its sinking further does not displace any more water.

But the familiar sort of phrasing I have used obscures what Stevin (and some years later, but independently, Galileo) discovered and that Archimedes and everyone else for 1800 years had missed: The weight of an object in water is reduced not by the weight of the water that is actually displaced but by the weight of the water that could be displaced if it had occupied the immersed space. For it is possible that the space available for a volume of water plus the immersed volume might together be barely larger than the immersed volume alone. If the object is less dense than water, it still floats! A very large block of wood can be kept afloat by a very small volume of water. That is what we now call the *hydrostatic paradox*.

This discovery seems hard to get without some sort of around-the-corner move. We call it the hydrostatic paradox just because it is so highly counterintuitive; in fact, as Galileo put it, it is "almost incredible." Archimedes missed it, and so did everyone else for 1800 years. But two early Copernicans saw it c. 1600.

The logic itself is not terribly difficult. If the weight of a body is diminished under water by just the weight of the water displaced, where does the weight go? If the surrounding water drains away, the body now regains the weight. Where does it come back from? Apparently there must be a force pushing up (supporting the body) that is just equal to the weight of the water that is pushed out of the way by the submerging object. So sinking into the water is like pushing down on the spring in a scale. As a body sinks into the water, it creates a spring pushing back equal to the weight of the water it has pushed out of the way. A body that is less dense than water, therefore, eventually makes the compressed spring push back with a force that balances the force of gravity drawing the body down—so it sinks no further. It floats. And if the body is denser than water, it never reaches this balance; it just keeps sinking, but it weighs less under water by the weight of the water displaced. But what is this watery spring?

Galileo and Stevin both worked this out with the help of oddly contrived arrangements that could bring out aspects of this hidden spring beyond those routinely encountered with overflowing baths, lifting anchors, and so on. Eventually each was prompted to the surprising conjecture that the net upward force doesn't depend on the size of the body of water. On the other hand, can't anyone see that a cork floats no higher in the water in a pond than it does in a cup? But as with Stevin's balls on a slanting surface and Galileo's free fall, the connection becomes too obvious to miss only after someone has made it too obvious to miss. Someone must create a situation in which what is occurring around us all the time appears in such a striking form that it *becomes* impossible to miss.

It is hard to extrapolate from commonplace experience to guessing that the floating body could still float even if the mass of water it had to float in were only one-tenth or one-hundredth or one-thousandth the mass of what is floating. Once the conjecture has been made, all it takes to confirm it is to confine a floatable object where it could be visibly supported by a narrow surrounding volume of water. That would not be an AC move. But the conjecture is so odd that it is hardly likely to occur even to a Stevin or a Galileo until *after* an AC move has brought it into sight. (Both write about it as something that still astonishes them.) Apparently each was led by thinking hard about what I've called the apparent "spring of water" to odd (around-the-corner) trials that resolved this puzzle in such an unexpected way.

Stevin's arrangement (Figure 5-2), on the other hand, is so odd that it does not look like a tool of discovery (in his book he first reports other odd but not *so* odd arrangements). It looks instead like a demonstration arranged to convince his readers, parallel to his at-a-glance demonstration in Figure 5-1. But the arrangement again has a definite AC character. No one would directly see this as a way to demonstrate the paradoxical effect. Stevin's setup here looks quite

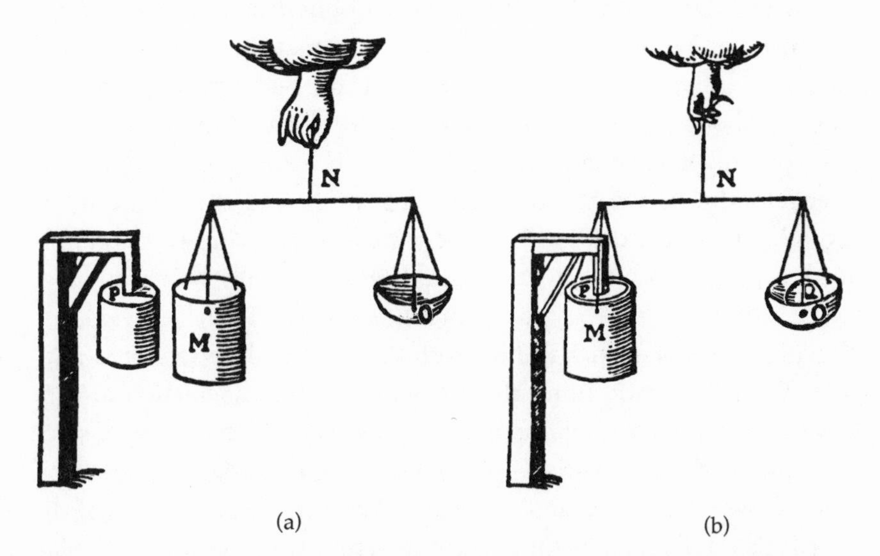

Figure 5-2 Stevin's demonstration of the hydrostatic paradox. (*a*) A full pail of water is balanced against the weight on the pan. (*b*) The pail is empty but is placed where it surrounds a plug that is slightly smaller than the pail. A small volume of water poured around the plug will then lift the far heavier weight.

as loony as the Cardano balance shown in Figure 3-1. His claim that adding a little water to the pail will lift a 10-pound weight on the other side of the balance sounds about as implausible as the claim that if you are really careful (and make sure that the bottom of the pail is rounded), Cardano's arrangement will not topple over. But unlike Cardano's arrangement, Stevin's works! A pound of water added to the pail on the left will in fact lift the 10-pound weight on the right. All the hydraulic machinery (presses, brakes, etc.) we now routinely use exploits this principle.

Gilbert

Around the time, or not long after the time, that Stevin was producing the AC performances of Figures 5-1 and 5-2 in the Netherlands, across the Channel, William Gilbert was beginning his experiments with magnets.

Since Gilbert took care that his papers would be preserved, depositing them with the Royal College of Physicians, we might hope to know something about how he came to that. But everything was then lost in the Great Fire of London in 1666. So we end up knowing nothing about how his work developed when (had it not been for the fire) we might have known a great deal. In his famous book on the magnet—the very first book reporting an extended and ambitious experimental investigation[3]—Gilbert mentions Stevin's work on navigation (twice), which is what would most naturally be mentioned, since Gilbert's book is itself much concerned with navigation. Stevin's books on statics and hydrostatics remained untranslated until after Gilbert's death. But Gilbert's boldness and imagination in contriving experiments might have been indebted to Stevin anyway, since it would hardly be surprising if some word of the striking demonstrations shown here in Figures 5-1 and 5-2 had made its way across the North Sea. Influence in the reverse direction is not in doubt; Stevin, Kepler, and Galileo all wrote about their admiration for Gilbert.

Gilbert came from a well-to-do family. He was educated at Cambridge, spent four years on the Grand Tour, then established his medical practice in London. He eventfully became personal physician to Queen Elizabeth, then to James I after Elizabeth's death. At the time of his own death in the plague year of 1603, he was president of the Royal College of Physicians.

Gilbert had the odd distinction of being a William with a younger brother who was also named William, the son of a late second marriage. It was the second William who eventually arranged for printing of Gilbert's *De Mundo*, which had circulated only in manuscript during his lifetime.

Given the English national interest in navigation, it is hardly unlikely that someone active in the English court might become interested in magnets and compasses. A retired seaman and compass maker named Robert Norman had published a little book on magnets and navigation in 1581. This was Gilbert's point of departure. But the book that Gilbert published in 1600 was far bolder and broader than Norman's. Norman in turn had proceeded from a few pages of notes on experiments with magnets three centuries earlier. This prior work attracted attention; at least twenty-eight manuscript copies survive.[4] On the other hand, when finally printed in the mid-sixteenth century, it was sufficiently obscure that a plagiarist attempted to pass it off as his own. Peter illustrates very well the point that doing experiments per se could not be the key to the burst of discovery that started c. 1600. Not only did he do careful experiments—though only a handful compared to Gilbert—but he long anticipated Gilbert's explicit commitment to the importance of doing experiments: He tells his reader that skillful experiments "will be able to correct errors which in a lifetime a person could never correct by reason or mathematics." Peter and then Norman come excruciatingly close to Gilbert's great discovery that the Earth itself is a giant magnet. But then again, Archimedes comes close to seeing the hydrostatic paradox. As with many other things, the breakthrough step was left for a Copernican to make c. 1600.

As mentioned in the Introduction, Gilbert avoids *saying* that he is a Copernican, but he is not shy at all about *showing* that he is Copernican. Figure 5-3, from his *De Mundo*, is Gilbert's picture of

Figure 5-3 The world system as presented by Gilbert in his *De Mundo*. This is from the printed version of 1651, but the manuscript version circulated before his death is essentially identical. It can be compared with the diagrams by Tycho (Figure 2-1) and Digges (Figure 2-6).

the world system. It is almost identical to Thomas Digges's explicitly Copernican diagram (Figure 2-5). Neither Gilbert nor Digges drew an explicit path for the Earth, but their diagrams are easily contrasted with Tycho's (Figure 2-1). In Tycho, the Sun (accompanied by the orbits of the planets) explicitly orbits the Earth, and the diagram clearly leaves room for that. In Digges or Gilbert, if the Sun were to orbit the Earth, it and all the planets with it would crash into the region of the fixed stars. And if you look at the sampling of remarks from Gilbert,[5] you will see that he repeatedly says things that are individually odd, and in the aggregate completely bizarre, if he is *not* a Copernican. I leave also to the notes some comment on why Gilbert might have been so cautious in what he explicitly says about the Earth's orbital motion. But recall that in the year Gilbert's book was published, the mystically Copernican Giordano Bruno was burned at the stake in Rome.[6] A few years later, in Protestant Holland and despite his close ties to the prince, Stevin was attacked for sacrilege when he published his Copernican book.

Of the experiments Gilbert reported in *De Magnete,* some are only replications of what Norman and Peter had done before him, and many others are direct elaborations of the earlier work. But others have a markedly around-the-corner look. In one, Gilbert arranged to have an exceptionally large lodestone marked at its north-facing end before it was dug out of a mine. At the surface, this 20-pound rock was then floated on a raft in a large tub. The raft dutifully rotated to leave the lodestone with its marked north end indeed pointing north.[7] Plainly this digging-plus-floating was not something that someone just fooling around with magnets would just happen to do, or see as an obvious thing to do. It can't come from tinkering with whatever is already at the focus of attention, but rather seems to require thought about where, *other than* the obvious place to look, evidence might be found. It seems to require a person alert to the possibility that powerful evidence might be around the corner, where you won't find it unless you think hard about where you might profitably look.

Gilbert's most important experiments turned on elaborate work with a piece of lodestone shaped into a sphere. Peter Peregrinus had used that arrangement long before. Shaping a lodestone into a sphere makes it easy to locate the poles. Let a nail fall, marking its direction. Repeat that at another location. The intersection of great circles along these lines locates the north and south poles. Peter showed ingenuity, but his work involved direct tests. Anyone who is familiar with magnets knows that they have poles, but it is hard to locate the poles on an irregular lodestone. However, if the lodestone has a conveniently round shape, then it becomes easy.

Gilbert did something much bolder. For Gilbert, the rounded lodestone became the *terrella*, a "little Earth." Gilbert reviews many details of what navigators have noticed about the behavior of compass needles contingent on where they are on the globe. No one could observe the interior of the Earth, but obliquely Gilbert argued that something about what is inaccessibly deep inside the Earth can be learned by experimenting on his *terrella*. He made his little Earth a little irregular, to simulate the influence of continents on the full-scale Earth. In many subtle ways, Gilbert showed that the little Earth and the big Earth exhibited identical properties. Beneath the complex surface of the Earth, there is presumably some substance that forms its core. If this substance, which is out of our reach, exhibits everywhere properties exactly like those of lodestone and contrasting with the behavior of any other thing we know, then what would a reasonable person believe but that the Earth at its core is itself a giant magnet?[8]

Gilbert in fact was not exactly right in his claim that the Earth's core is lodestone. The core of the Earth is iron, not lodestone (a mineral rich in iron). But as Galileo later said, Gilbert did not get everything right, but he got things started, enabling others to go even further.

No one had ever made as heroic an argument as Gilbert's, with the one exception that makes sense here: Copernicus himself. We have what very quickly became the classic example of bold around-the-corner inquiry [AC Case ii], in the inverted form allowed for in the taxonomy of Figure 4-2. Gilbert wants to understand the behavior of compass needles as mariners sail the globe (*B*). That must turn on some source of attraction (*A*) that is wholly out of reach. Before Gilbert, that source of attraction was thought to be in the stars, or perhaps in mountains of iron at the poles. But it had been known since Peter Peregrinus that needles falling on a rounded lodestone point to its poles, so perhaps what attracts the mariner's compass needle is the Earth itself. That seems simple to us. But no one before Gilbert had thought of it, which argues that until you come to believe it, the idea that the Earth is a giant magnet sounds very implausible.

You can't manipulate the Earth, or dig down far enough to reach its core. But Gilbert plainly thought hard about what might be noticed from tests on a tiny model of the Earth (C) to see what happens (*D, D′, D″*, etc.): AC Case ii in Figure 4-2. And with extensive enough agreement—and Gilbert shows very extensive agreement—between what happens with slivers and such on his *terrella* and what mariners report happens with their compasses, we might reach a conclusion. Bacon thought that that was nonsense, even after reading Gilbert.[9] But Galileo has his interlocutors in his *Dialogue* agree that Gilbert's argument that the Earth is itself a magnet is "as convincing as is permitted by natural science" where "geometrical certainty cannot be demanded."[10]

Gilbert's mining and then floating of his twenty-pound lodestone patently needs an argument to explain why such an odd performance is relevant to whether the Earth is a magnet, just as Stevin's loony-seeming balance needed an argument to make sense of why someone would do that. But following an argument, of course, is much easier than discovering one. And once insight or evidence that had been around-the-corner is pulled into view, it is of course no longer around-the-corner.

Hence (again) the "purloined letter" aspect of around-the-corner inquiry comes into play. Once obliquely located evidence or an unexpectedly relevant argument is brought into sight and becomes familiar, there is no longer anything around-the-corner about it. Around-the-corner inquiry has a self-liquidating aspect. AC inquiry was never seen as an innovation of method, even when, around 1600, it appears to have been brand new. After all, no one ever supposed that an investigator should *not* try to find a novel argument or novel evidence around the corner. It just had not occurred to investigators to put in the effort required to notice where a promising corner might be. Before AC inquiry was in place, no one missed it. Afterward, no one noticed that it was now there!

Galileo

Galileo eventually told his readers that Gilbert would be persuasive to "every man who has attentively read his book and carried out his experiments." The contrast is with people who don't listen carefully to arguments and don't attend carefully to experiments. But in the course of explaining why he found Gilbert so persuasive, Galileo offered a more extended view of Gilbert's method. "Gilbert's book," Galileo has Salviati say, "might never have come into my hands if a famous Peripatetic philosopher had not made a present of it, I think in order to protect his library from its contagion."[11] Gilbert's method, Galileo says (through Salviati), has "a certain likeness to my own."[12] The key thing, as Galileo presents this method, is what William Whewell much later would call "consilience," or what I will call simply "fit," but which in the taxonomy of Figure 4-2 is presented as around-the-corner inquiry Case ii.

And Galileo appears to have been much more than merely gracious in praising Gilbert's method. While Gilbert's papers were destroyed in the Great Fire of London, so that we can say very little about the development of his work, Galileo's papers survive, and (as we've already seen in Chapter 4) the careful work of Stillman Drake and others gives us unexpected detail on how Galileo proceeded. He seems to have been aground in his effort to understand motion as of about 1590. He had a treatise in hand, in two versions, but in the end it was not work he wanted to publish.[13] Abandoning what he had plainly worked hard to produce makes no sense for an ambitious young scholar unless he knew that there was something wrong with it—as indeed there was, but for a dozen years after 1590 Galileo made no noticeable progress. Then, in 1602, he read Gilbert. Gilbert died in 1603, three years after his book came out. Galileo was then nearly forty. But for the Galileo who made his mark in history, his scientific life only then was beginning.

After reading Gilbert, Galileo quite abruptly knew how to make discoveries. Over the next half dozen years, he produced most of the discoveries that would be formally published many years later in his *Dialogue* (1632) and *Two New Sciences* (1638). These results routinely came from his careful (often geometrical) reasoning interacting with AC moves that yielded ingenious but often simple experiments.

So although there was no way for Galileo to *directly* observe free fall with the precision he needed, he was eventually able to find his way by the sort of cleverly contrived AC inquiry discussed earlier. And for the rest of his career, Galileo was a virtuoso at that. What no one could see directly, he could somehow ferret out with arguments and tests of a Gilbertian sort: Look over here, where we can see clearly what is happening, which makes sense of what must be happening over there where our attention was fixed.

Galileo rediscovered Stevin's hydrostatic paradox, prompted by a debate, carried out partly for the amusement of the Florentine court, over what determines whether a body will float or sink. He was defending the view of Archimedes that a body sinks or floats depending on whether it is more or less dense than water. His Aristotelian adversaries challenged him to explain why an ebony chip, which is denser than water, would nevertheless float.

Galileo noticed (which no one before had, although everyone has seen slivers of metal float) that the ebony chip actually sank slightly into the water, with a little wall of water around its perimeter (due, we now would say, to surface tension). Somehow, as with the much more familiar example of water drops, water has some tendency to hold itself together rather than spread out. Therefore, the weight of the total volume (the sliver of ebony sunk into the water plus the resulting air pocket inside the little wall of water) in fact matched the weight of water displaced, as Archimedes said it should. If you wet the chip (Galileo showed, looking around a modest corner), then there is no longer a boundary between the water and the dry chip for the wall to form along, and the same chip that had previously floated will now sink. Since the wall cannot be sustained beyond a small height (water drops can get only so big before the surface tension is broken), the chip has to be thin. Galileo showed that the effect is even more noticeable if a bit of gold foil is the floating object. The gold is very dense but can be beaten out to be very thin.

Although it is clear to us today that Galileo was right and his adversaries were wrong, it was not so clear then. This was a problem that direct inquiry could not completely resolve because the Aristotelians could show that the relationship between density and floating was more ambiguous than Archimedes had noticed. Usually, of course, things that are denser than water sink, and only lighter things float. But the ebony chip, which is denser than water, nevertheless floats, which left it doubtful that Archimedes (and Galileo) had actually gotten things right. So Galileo was prompted to try to see more. And indeed he saw much more, since this led him into more general puzzles about floating bodies and eventually to an account of the "almost incredible" hydrostatic paradox.

Kepler

So far we have noticed around-the-corner moves in the work from Stevin as far back as the 1580s, and then on a larger scale in the book Gilbert published in 1600 (in Latin, so it was accessible to Stevin, Kepler, and Galileo, all of whom wrote admiringly about it). And then (but only then) we find more such moves in Galileo, and deployed over a wider range of topics. Given this timing, with Galileo apparently stuck for a dozen years until he read Gilbert, and with his explicit tie of his own method to Gilbert's, it looks as if Gilbert's example influenced Galileo. And (more speculative) Gilbert himself may have been influenced by Stevin, given the proximity and mutual interests of the Dutch and English courts. Kepler's resolution of the often discussed (since Aristotle!) but never clearly resolved puzzle of how images are formed through small apertures also comes very soon after Gilbert's book appeared. We can't be sure whether Kepler had read Gilbert before he made this discovery using the striking AC move to be discussed shortly. But that he read Gilbert and was enormously impressed, there is no doubt. In the book reporting on this optical work (published in 1604, but sent to press a year earlier), Kepler announced his hope to cross the North Sea and meet Gilbert. He hoped to win Gilbert's friendship, Kepler writes, "through my eagerness to learn."[14]

Kepler came from modest parents, and on his own account very disagreeable parents, both with each other and with outsiders.[15] He was a sickly child, but, managing to survive both ill health and unfortunate parents, he got a good education on scholarships intended to produce clergymen. He was diverted from that vocation when his teachers recommended him for a vacant post in mathematics. At the time, being a mathematician entailed responsibility for teaching astronomy (and astrology), so, like Galileo in a parallel situation, he necessarily became acquainted with Copernicus, and also like Galileo, Gilbert, and Stevin—and in contrast to most of his contemporaries—he came to believe Copernicus was right.

Kepler's ardently Copernican first book came out in 1596, as described in Chapter 2. He sent two copies with a traveler to Padua, to give to likely readers at that city's famous university. Both copies were given to Galileo, which suggests something about the number of likely readers of such a book. Galileo promptly wrote to Kepler saying that he agreed with him and would say so in public if he was not surrounded by so many fools.[16] This has sometimes been read as showing the contrast between the devious Galileo and the forthright Kepler. Kepler indeed was very forthright, but Galileo was probably only being prudent. He was an untenured junior faculty member on a modest salary, with a mistress and three children to support and dowries to pay for his sisters. There was much more tolerance, if not much more support, for Copernicus in Germany, where Copernicus was thought of as German.

By the time Kepler arrived to work with Tycho in 1600, Tycho had had a row with his patron's successor and had moved his operation from his "heavenly castle" off the coast of Denmark to a castle outside Prague provided by the Holy Roman Emperor. When Tycho died a year later, Kepler succeeded his employer as chief mathematician (meaning, mainly, astronomer and astrologer), but he was not rewarded on the same grand scale as the noble Tycho. He was not given a castle. A strange claim of Tycho's (that he observed the Sun to shrink during an eclipse) then led Kepler into work on the puzzles of images seen through apertures. Kepler's remark about wanting to meet Gilbert comes from the report on this work, which was published in 1604, but was submitted to the court in 1603 to show that Kepler was earning his salary.

Kepler's "Book" Experiment

Kepler eventually decided that Tycho's report of a shrinking Sun during an eclipse was only a blunder, that Tycho had just not been careful enough in handling his instruments. But just as Galileo was prompted by the Aristotelian challenge about the floating chip of ebony to broader work that turned up the hydrostatic paradox, Kepler was led from Tycho's odd claim to the famous related problem already mentioned here, that (for example) during an eclipse of the Sun, light filtering through gaps in foliage will form images on the ground showing perfectly the progress of the eclipse. Each gap in the overhanging foliage has a unique shape, but the light pattern on the ground through each opening shows the identical shape of the Sun.

So the puzzle is, how does light passing through the heterogeneous and irregular gaps in the overhanging leaves form the multitude of identical images that always match the actual shape of the eclipsing Sun? This is the problem of pinhole images, of which various other naturally occurring examples had been noticed since ancient times. The puzzle had been attacked by all leading writers on optics for a long time, but it was left to Kepler solve it. And (perhaps only because of the almost compulsively confessional nature of his reports) we happen to know just how he did it.

Kepler resolved the puzzle by an analysis which in contemporary language came down to this: Divide the luminous body (say, the Sun or the illuminated Moon) into arbitrarily small pixels. Each pixel, however small, irradiates the entire aperture, and the rays from the pixel go in straight lines to the illuminated surface. This much Alhazen in the twelfth century and before him Ptolemy in the second century had recognized.

These straight-line rays from the pixel, consequently, must light up an area beyond the aperture that plainly must have the shape of the aperture. But that illumination would be faint, since the light from any tiny pixel will be spread thinly across an area that is very large compared to the pixel.

Therefore, the image from each pixel is the shape of the aperture—for example, a triangle for a triangular aperture. But the image we see is not that of any single pixel shining thinly through a triangular aperture; it is the aggregate bright image produced from a vast number of faint triangles. Together, the aggregate of numerous faint triangular images accurately reproduces the bright shape of the luminous body!

It takes a complicated geometrical analysis to pin down the conclusion that the net effect of a vast number of faint triangles will be a well-formed crescent (for the case of a partially eclipsed sun). Perhaps inspired by a technique Dürer used in training pupils in perspective drawing (Figure 5-4), Kepler performed an experiment that might have been undertaken by any of his predecessors over the 2000 years this puzzle of pinhole images had been discussed. As he reports:[17]

"I placed a book on high to take the place of the shining body. Between it and the floor I set a tablet containing a many-cornered aperture. Next, a thread was sent down from one corner of the book through the aperture and onto the floor. It fell to the floor in such a way that it grazed the edges of the aperture. And I traced out the path produced and so created a figure on the floor similar to the

Figure 5-4 Dürer published this engraving in 1525. It shows the gadgetry that he devised to train pupils in perspective drawing, which may have suggested to Kepler using threads to represent light rays in the experiment he used to resolve the puzzle of pinhole images.

aperture. Likewise, by means of a thread attached to another, a third, a fourth corner of the book and finally to an infinite number of points along the edge, there resulted on the floor an infinite number of traced figures [each having the shape] of the aperture, but which together produced a great and four-cornered [figure having the] shape of the book."

Kepler went on to produce the first workable theory of how images are formed in the eye and then more generally behind lenses, starting with water-filled spheres to simulate the aqueous humor of the eye. He thus was the first to realize that the images in our eyes are upside down. This was extremely puzzling to Kepler, as indeed it is to a modern reader when first encountered. With his usual candor, Kepler presented the result as bewildering but somehow true anyway. Generalizing this led him to a wider theory of how images are formed behind lenses. With this theory in hand, when Galileo's report on the telescope was published (a half dozen years later, as is discussed later in this chapter), Kepler in a matter of weeks worked out a theory of how telescopes work, including a proposal for an alternative design that performed so well that it eventually displaced Galileo's. The book Kepler published on this (*Dioptrics*) became the foundation on which Newton wrote his own famous *Optics*.

Kepler's Laws

The most famous items in Table I-1 are Galileo's discoveries with the telescope and Kepler's laws. But was either contingent on AC inquiry? Kepler exploited Tycho's observations, and Galileo exploited an invention made by Dutch lens makers. Given Tycho's observations, what AC move was needed for Kepler to work out his laws? Given the basic idea of a telescope, what oblique move was needed for Galileo to turn an improved version of that instrument on the heavens?[18] But consider what Kepler and Galileo actually did.

For all the discoverers of Table I-1, the motions of the planets must be somehow driven by a force emanating from the huge and central Copernican Sun. But this point was especially critical for Kepler, who was not only Copernican, but more Copernican than Copernicus himself was—or indeed could plausibly have been. Kepler is guided throughout by his belief that astronomy ought to be rooted in a *physics* of the heavens, deriving somehow from the Copernican revelation. With quite overwhelming perseverance, he kept looking for ways to put that conviction to work, and the outcome of his "war with Mars" (exploiting Tycho's particularly detailed observations of Mars) was what he very reasonably announced as a "new astronomy."

Copernicus had provided a radical astronomy that was nevertheless technically Ptolemaic, as described in Chapter 1. Kepler provided a more accurate astronomy than Ptolemy (which Copernicus did not either claim to do or actually do). He abandoned Ptolemy wholesale, which also meant abandoning the great bulk of the technical work by Copernicus, which is essentially always Ptolemaic.[19]

But from beginning to end, Kepler saw his work as built on the fundamental Copernican insight that astronomy makes more sense if you see the Earth as a planet. When he eventually wrote a nontechnical summary of his work, Kepler called it *An Epitome of Copernican Astronomy*.

Kepler's abandonment of the venerable astronomical dogma of uniform circular motion has a heroic character. Very few astronomers could accept Kepler's ellipses until Newton had derived the full set of his planetary laws from more general laws of motion.[20] Kepler anticipated the delay: "I yield to the sacred frenzy. . . . I shall endure. . . . [I] can wait a century for a reader, as God himself has waited six thousand years for a witness!"[21] This sounds like the outburst of a madman. But it turned out to be just the simple truth of the matter.

Kepler's conviction that proper laws of planetary motion should reflect the central physical role of the Copernican Sun turned out to be strong enough to overcome what everyone else, starting long before Ptolemy, had taken to be beyond doubt: that uniform circular motion was the only physically plausible basis for astronomical models. Like Ptolemy, Copernicus endorsed that, and even Kepler only very reluctantly gave it up.[22] All were guided by physical intuitions and arguments that seemed irresistible, illustrating how what seems impossible to be wrong can turn out to seem not compelling at all once the habits of mind that go with the intuition have been broken. What changed here was that after Newton it could no longer seem that circular motion was still the most convenient device, which fit the mathematical methods that astronomers routinely used, which worked, and which seemed to have no alternative. All that turned out to be wrong, and while it took time, eventually the circular motion intuition that so long had seemed irrefutable faded into nothing.

But even while the inevitability of circular motion was still in place for Kepler, he also had the rival, intrinsically Copernican intuition that the immense Sun must provide the motive force that drives the planets; hence the variation in a planet's motion along its orbit must somehow depend on the Sun. Long before he could break the commitment to circular motion, Kepler felt sure that planetary motion could not be controlled from the empty equant point that governs Ptolemy's motions and in a slightly hidden way Copernicus's as well. Behind Ptolemy's mathematical fiction of motion guided from an empty point removed from the center, for Kepler there must lie a physically real force emanating from the Sun.[23] This led Kepler into an AC move, Case i. A Copernican world suggests that a force from the Sun moves the planets ($A \rightarrow B$). How to check that is certainly not obvious. But if it is right, there should be a rule that relates planetary speed to distance from the Sun: the closer the planet, the greater the motive force on it and hence the faster it should move. Not at all easily, but eventually, Kepler found a pair of rules (what we call his first and second laws) that turned out to work better than Ptolemy's rule. So (Figure 4-2, Case i): $B \rightarrow B'$, although exactly what B' might be or how to find it was a puzzle. Certainly there was no way to directly look and see it. But with great persistence, B' was eventually found, and it worked.[24]

Galileo's Discoveries with the Telescope

Just as Kepler was publishing his *New Astronomy* (1609), Galileo heard of a device that could make distant things seem closer. Before another year was out, he had astonished Europe with news of mountains on the Moon and small planets orbiting Jupiter. The first reaction of his Aristotelian adversaries was to claim that it was a hoax. But within another year enough people had seen what Galileo claimed to have seen—most notably the priestly astronomers of the Collegio Romano—to end the controversy. Even an early version of Galileo's telescope had obvious value to a naval power like Venice, since a ship that could detect a mast on the horizon sooner than its adversary would have a clear advantage. Galileo presented his telescope to the Venetian Senate, which rewarded him by doubling his modest professorial salary.

But Galileo had more ambitious things in mind. Given the curvature of the Earth, there was not much military value in a more powerful telescope, and it was not at all clear that there would be anything to be seen in the heavens, either. Kepler (in an enthusiastic response to Galileo's book) said that he had considered the telescope but had mistakenly supposed that the opacity of the atmosphere made the project unpromising. But Kepler was already profoundly engaged in searching for Copernican insights in another way. So it was Galileo who was seized with the prospect that remarkable things might be seen in the heavens if you worked really hard at making a telescope good enough to see them. And he had some basis for boldly pursuing the matter. The discovery of the hydrostatic paradox lay a couple of years in the future, but by 1609 Galileo had already contrived his clever experiments pinning down the fundamentals of motion. Galileo was not the reckless egoist that he has been often portrayed to be. His enemies really hated him. But his friends (and, as Dava Sobel, has recently made clear, his daughter[25]) really loved him, and had reason to love him. However, he was certainly not burdened with undue modesty. And he was seized with the immodest possibility of doing something like what Copernicus himself had done: find a novel way of looking at the heavens for evidence that the Earth was in motion. And he was bold enough and smart enough and persistent enough to do it.

What Galileo found with his telescope was, of course, bound to be discovered sooner or later. With time, artisans would make better and better telescopes. Eventually these telescopes would be good enough to distinguish details on the Moon and elsewhere in the heavens, and it requires no around-the-corner idea to account for some curious fellow taking a look at the Moon.

But the history is a bit different, and in a way that fits the AC story. The telescopes that were available when Galileo first heard of the device were toys. A person who looked at the Moon with these instruments (as no doubt happened) would see nothing that was of any great interest. The very first printed announcement of the new device, which appeared a year before Galileo's observations, mentioned that it could make out stars that were invisible to the naked eye.

And indeed, with this novelty that could make far-off objects appear closer, why not look at things that were really far off? There is nothing around-the-corner in that. But since the report claimed no more than that stars that were too faint to be seen with the naked eye could now be seen, we can be sure that its resolution was too poor to see significant new detail in such obvious targets as the Moon. And making a telescope that was good enough to do much better was no small feat. Lens makers in Flanders and Holland had known of the telescope for twenty years before Galileo heard of it, and they had obviously had more experience in making and working with lenses than Galileo did. But Galileo was motivated to look where it did not occur to others that there was something to be found. The key seems to have been persistence in finding what to look for. Lenses were made for use in eyeglasses. When they were doubled to make a telescope, imperfections that were of no importance in a pair of spectacles had (of course) a magnified effect. But Galileo thought hard enough about what was limiting the effectiveness of a telescope to learn how to produce lenses that would yield an instrument five times more powerful than those anyone else had made. Within a matter of scarcely more than a few weeks, Galileo had discovered how to sift through many lenses to find the few that were suitable for magnifying that much without losing too much definition. But a prior requirement was to figure out what characteristics in a lens it was crucial to look for.

Galileo's inspiration was doubly Copernican. To a Copernican, the Earth is a planet, so contrary to Aristotelian expectations a Copernican might expect to see something striking with a good enough telescope. But Galileo was also following Copernicus in looking around-the-corner for information that remained out of sight of everyone else.

Such ideas had occurred to other Copernicans. Even without the help of a telescope, Gilbert confidently asserted that the Moon is like the Earth and that the indistinct blotches on the Moon are mountains and seas. And another Englishman, Thomas Harriot, actually had a fair claim to being the first to actually see detail on the Moon with the telescope. But Harriot saw only that the terminator (the border between light and dark) was ragged. He could not see the craters whose shadows accounted for that, whereas Galileo could see them clearly enough to quite accurately calculate their enormous height. Galileo's telescope was sufficiently superior to entirely dominate discussion. But it is a remarkable coincidence, though on the argument here not a surprising coincidence, that the only two men in Europe who were moved to quickly find a way to make a telescope good enough to see something strikingly new in the heavens had something salient in common. Galileo, of course, was a Copernican. And so was Harriot.[26]

The End of the Beginning

Now review a bit. In 1600, Gilbert published his elaborately experimental argument that the Earth is a giant magnet. In 1604, Kepler published his *Optics*. In 1605, a translation of Stevin's work on hydrostatics finally made that work available to people who did not read Dutch.[27] In 1609, Kepler published his report on his "war with Mars," treating, as no one had before, astronomy as a kind of physics. He makes that point twice in the very title: *New Astronomy Based on Causes: or, Celestial Physics.* A year later, Galileo's report on the telescope startled Europe. A year after that, there was Kepler's *Dioptrics*, which provided the first adequate account of how images are formed by lenses, including a treatment of the eye as a lens. And in 1612, there was Galileo's report on the controversy over floating bodies. The list here does not consider discoveries that Galileo had communicated to correspondents before 1612 but did not publish until years later, since we are at the moment interested only in what was reasonably available to influence others.

The body of work that was in print by 1612 clearly reveals that a true flood of startling discoveries appeared in the very first years of the seventeenth century, marking an irreversible shift in the capacity to make discoveries. Anyone who was moved to study this work would have been influenced by its around-the-corner moves, although once absorbed, the around-the-corner approach would seem simply the natural way to proceed. A person who was familiar with these turn of the seventeenth-century exemplars would find it natural, not revolutionary, to think hard and think boldly about where striking argument or evidence might be found to test ideas that seemed in some way promising, even though hard to believe.

The most important new idea of the next decade was certainly Harvey's claim for the circulation of the blood, which was made by 1619, though it was not put into print until 1628. Harvey could see no connection between the arteries and the veins. How could blood circulate if there was no connection between the two systems? Harvey had to conjecture that the connections were too small to be seen, as indeed they were until a microscope that was refined enough to detect the finest blood vessels became available many years later. Nevertheless, he was able to find good enough evidence in an oblique manner to convince essentially everyone, prompting Hobbes to remark on how rare it is for such a revolutionary idea to triumph during the discoverer's own lifetime. The most important evidence was provided by a distinctly around-the-corner move: Harvey carefully studied the slowed heartbeat of *dying* animals, allowing him to plausibly calculate the quantity of blood pumped at a normal rate, which in a short time exceeded the total amount of blood in the animal's body.

There is no evidence that Harvey had any special interest in astronomy, so he was perhaps unaware of the geocentric dependence on pretzel-shaped orbits that so keenly affected men like Kepler and Galileo. But unlike the turn-of-the-century discoverers, Harvey did not need a directly Copernican inspiration. He could follow in the footsteps of Copernicans like Gilbert, Galileo, and Kepler without being Copernican himself. And being a successor to Gilbert as physician to the king, Harvey had an especially salient model to follow.

I began this book by showing a table that demonstrates the stark contrast between the number of memorable discoveries in the years immediately around 1600 and the number during the previous thousand years and more. But I could have picked *any* interval after 1600. After 1600, it is always easy to point to memorable discoveries, and it is also always easy to point to examples of AC inquiry. In a narrow span of years bracketing 1600, three unprecedented conditions came together, as we have now noticed for each of the four most prominent figures. Of Stevin, Gilbert, Kepler, and Galileo we can say: (1) They were all ardent Copernicans; (2) they all used AC inquiry; and (3) they all made discoveries that were more striking than anything in the prior thousand years and more. And this makes a turning point worth noting. We might even say that it was revolutionary.

Notes

1. So the full weight is pulling (of course) when the angle is vertical ($\sin 90° = 1$), but only half the weight is felt when the angle is 30°. More balls are required to cover the less inclined—and hence longer—arm of the triangle.
2. Galileo amused an audience by exhibiting an object that sank halfway and then stopped. However, this was a trick. He filled a vessel halfway with seawater, then inserted a thin sheet, then added a layer of fresh water above it. By carefully removing the sheet, he left the bottom of the tank filled with denser water than the top. A wax ball weighted with lead filings could then sink through the fresh water but float on the salt water.
3. By comparison, earlier reports on experiments, including the earlier work on magnets that clearly gave Gilbert his start, are only pamphlets.
4. I take this and other details from Thompson, 1906.
5. See the notes to the Introduction.
6. Recent scholarship makes it seem unlikely that Bruno's Copernicanism was central to his fate. But it would hardly look that way from London, where Bruno had learned about Copernicanism, probably from Digges, around 1580. Of course Gilbert had nothing to fear from the Inquisition. But questioning the literal truth of scripture was not a trivial matter. A friend provided the introduction to Gilbert's *De Magnete*

mentioned in the notes to the Introduction. This assured readers that Gilbert did not go further than the Bible allowed.

7. *De Magnete*, p. 120. Gilbert's point is that the north-facing end of a magnet near the surface of the Earth (so that it is separated from the magnetic core of the Earth) will be the south pole; hence, when floated, this end will be drawn toward the north. He shows with the *terrella* that if a bit is dug out of the body of the magnet, the reverse will hold: The end that had been to the north will turn to the south.
8. My language here paraphrases Galileo's *Dialogue*, pp. 400ff. See the discussion later in the chapter of Galileo's comments on Gilbert.
9. Bacon, an eloquent advocate of induction from experience, won attention among actual scientists later in the seventeenth century that contrasts with what was never better than indifference from those who made the discoveries that coincide with the onset of the new century. And Bacon thought no more of these discoverers than they thought of him.

   While Bacon was alive, those who are remembered now for their discoveries proceeded without noticing Bacon. And while Bacon wrote much about what he called experiment, when he noticed what pepple like Gilbert and Galileo were doing, it was mostly with disapproval. What they were up to was not what Bacon intended. It is a historical oddity that Bacon became ex post the prophet of modern science. Bacon's comments are mostly on his fellow member of Elizabeth's court, Gilbert. He did not like Gilbert's work very much. In particular, he thought Gilbert's claim that the Earth is a magnet to be absurd. See Suzanne Kelly, *The De Mundo of Gilbert* (Menno Hertzberger, 1965).
10. *Dialogue*, p. 410.
11. The quotes here are from the closing pages of Day 3 of Galileo's *Dialogue*, pp. 400ff.
12. *Dialogue*, p. 410. The remark is made by Sagredo to Simplicio. Sagredo is the third party in the *Dialogue*, who often judiciously sums up after discussion between Simplicio (the Aristotelian) and Salviati (speaking ordinarily for Galileo himself). It is Salviati (p. 403) who describes Gilbert's method as "like my own."
13. Galileo wrote out his argument both as an expository treatise and in the form of a dialogue.
14. From Kepler's *Optics* (1604), 2000, p. 148.
15. His father, Kepler wrote, was "vicious, inflexible, quarrelsome and doomed to a bad end," which very likely he had, but since he aban-

doned his family, no one knows. Kepler's mother (as he described her) was "small, thin, swarthy, gossiping and quarrelsome" (quoted in Koestler, pp. 22–23). At one point, Kepler had to interrupt his work to return home to defend his mother, who was being tried as a witch.

16. Kepler remarked to a friend that he had heard from an Italian whose first name was the same as his last (that is, Galileo Galilei).

17. Straker, p. 390, as quoted in Lindberg, 1985. Note that what Dürer himself did would *not* be an example of oblique inquiry. That an image in correct perspective could be formed by treating an object "as if" straight rays ran from the object to the eye of the viewer had been the familiar (and correct) view since classical times. But as discussed earlier, this worked practically, but why it worked was a mystery. Dürer's illustration of his training method was available for seventy-five years before a Copernican realized that something along these lines could be used to resolve the famous problem of images. Dürer's gadget itself involved only directly helping a student by running physical strings along the pencil lines that had been shown for centuries in drawings made to illustrate perspective. But Kepler's book took the place of a luminous object, and what he traced out with his use of strings was not the "as if" single ray from each point converging to the eye but the perimeter of an infinitely large number of rays that reach every possible point.

18. Here the moons of Jupiter were especially important. An awkward point for the Copernicans was that if the Earth is only a planet, how could it hold onto the Moon? This necessarily faded as an objection as conservative astronomers gave up on Ptolemy (Chapter 2), since in Tycho's alternative five orbits accompany the Sun as it orbits the Earth. But finding a system much like the Copernican system of the planets in the moons of Jupiter was still potent, as were many other details revealed by the telescope (sunspots, solar rotation, mountains on the Moon, etc.) that were readily compatible with a Copernican world but puzzling in a geocentric world.

19. We remember what turned out to be profoundly important. But although Kepler certainly claimed victory in his war with Mars, and boldly applied what worked for Mars to the rest of the planets with great success, it is surprisingly difficult to find a clear discussion of his laws in Kepler's own writings. Kepler himself seems to have been at least as enthusiastic about a discovery that we would just as soon forget. He always thought he had made a terrific discovery when he fit the heliocentric orbits of the planets to a nesting of the five Platonic solids. The arrangement does not work for the Ptolemaic system, and it does

not make much sense for the Tychonic. Before the century was out, Herschel had discovered a planet unknown to Kepler (Uranus), so that much as this idea pleased Kepler, his proof that there should be exactly six planets (to accommodate the spacing provided by the five Platonic solids) provides a nice illustration of a beautiful theory destroyed by a nasty little fact. But, right or wrong, it was certainly around-the-corner.

20. (1) Planetary orbits are ellipses; (2) as they move around the Sun, the planets sweep out equal areas in equal times (which makes the planets go faster when nearer the Sun); and (3) the periods of the planets scale with the 3/2 power of their mean distances.

21. Quoted in Gingerich, 1993, p. 47. The context was Kepler's discovery of the last of his three laws.

22. Ptolemy's use of the equant motion here raised the puzzle already mentioned in Chapter 1. On this, Kepler followed Ptolemy, not Copernicus; indeed, he eventually found the physically satisfying foundation that he sought for the effectiveness of the equant motion. But since the solution involved abandonment of the axiom, even Kepler came to it only with great difficulty.

23. Gingerich, 1983, gives a simple demonstration of why Ptolemy's rule works so well. Ptolemy's planets sweep out equal angles as seen from what we would today describe as the empty focus of an almost circular ellipse, which turns out to closely approximate the effects of Kepler's laws (equal areas as seen from the occupied focus of the ellipse). As mentioned in Chapter 2, Ptolemy does not attempt to give a physical interpretation of how that might happen. Copernicus and (in a Ptolemaic framework) Arab astronomers before him avoid this physical anomaly by assuming an extra little epicycle whose effect is to closely approximate Ptolemy's equant motion. Kepler always ignored this aspect of Copernican astronomy, in contrast to Tycho and other geocentric astronomers, who admired it greatly. In terms of what Kepler was seeking, it looked like the blind alley it turned out to be.

24. Kepler got part way quickly. In his 1596 book, even before he had Tycho's data, he already made the radial velocity of the planets depend on distance from the Sun. But the more effective version that we remember came only with success in his "war with Mars."

25. Dava Sobel, *Galileo's Daughter* (1999).

26. Along with a prize student, Harriot set eagerly to studying Kepler's *New Astronomy*. But as anyone who has looked at Kepler's book will know, this is not a text anyone would be eagerly studying if the reader was in doubt that its fundamental premise was right. Harriot's student remarked that he was dissatisfied with Copernicus's circular astronomy. But unless he was a Copernican, why would Harriot be dissatisfied with the circular aspect of Copernicus's astronomy: In this respect Copernicus is no different from Tycho or Ptolemy.

27. The appearance of Stevin's discovery of the hydrostatic paradox in a more accessible language than Dutch may account for the absence of a translation from Galileo's vernacular Italian of his own work on "floating bodies" of 1612 (discussed below). Galileo's main novelty was his version of the hydrostatic paradox, by then already available (as of 1605) from Stevin.

CHAPTER 6

The Emergence of Probability

Here is something that by now is a familiar story, but delayed by half a century. The sudden development of probability theory has been a puzzle for historians. The basic principles of a calculus of probability were logically within easy reach of the mathematicians of the ancient world, yet we do not see the theory emerge until the 1650s. Gambling, including gambling with symmetrical dice, was common from very ancient times. As a mathematical problem, the breakthrough to a calculus of probability is hardly more difficult than, say, proving the Pythagorean theorem. Why should one have come 2000 years behind the other?

We have a case of a mathematical discovery in which the actual mathematics of the breakthrough are so simple that we can march through the logic in a few paragraphs, making no demand on a reader much beyond knowing how to count. In fact, we will have occasion to do that twice: once the way Pascal did it, and once the way Fermat did it. That it took 2000 years for this discovery to appear argues powerfully that some now-invisible step was required for mathematicians to finally come to see what seems so obvious today. And it turns out that there is a direct line of AC moves running from Stevin and Galileo's discoveries near 1600 to this mathematical discovery of Pascal and Fermat near 1660. I start by reviewing that line of descent.

Does Air Have Weight?

By analogy with what he had learned about water pressure—the "across-the-street" variant of AC (Figure 4-2, Case ii)—Galileo could expect that a parcel of air *in air* effectively weighs nothing, just as a bucketful of water weighs nothing underwater. Lifting the bucket (slowly[1]) is really easy—until the bucket comes out of the water, at which point it becomes heavy. But if you could cram several buckets of water into one bucket, then it would be easy to feel the weight of water even under water. You can't do that with water, since it is incompressible. But a person who got this far might look for a way to cram multiple volumes of air into one volume of artificially dense air.

Galileo found a measurable difference between an "empty" jar (containing a normal volume of air) and the same jar when filled with as much air as he could compress into it. He then opened a valve to let the air in the jar return to normal pressure, capturing the surplus by letting it bubble into an upside-down vessel of water. The difference between the two weighings gave him the weight of the volume of air that had been crammed into the jar. The volume of air at the top of his collecting bottle showed him what that volume was.[2] This combination of moves let Galileo establish that air did indeed have weight.[3] And Galileo had also learned a superficially unrelated point: water can be pumped only up to about thirty-four feet. Here the information came not from experiments, but from his plumber.[4]

The next step came from Evangelista Torricelli, a young mathematician who had come to learn what he could from Galileo and soon was living at the country house outside Florence where Galileo was (technically, and in some ways literally) a prisoner of the Inquisition. Eventually Torricelli succeeded Galileo as chief mathematician to the duke of Florence. Torricelli had what Galileo (in his comment on Gilbert's method) called a lucky thought, and a very famous lucky thought. It occurred to Torricelli that just as a fish could not know of the weight of the ocean above it, we may be living at the bottom of an ocean of air without noticing the consequences of that.

The origin of this flash of insight (the first epiphany) was apparently just that since air had weight, more air must have more weight. This hardly seems a stunning insight. But since our experience is with air in air, which effectively weighs nothing, it is not easy to notice. Putting a brick on a box will put pressure on the box, and adding bricks will add more pressure, until the box collapses. It occurred to Torricelli that all the air piled up over the surface of the Earth must put pressure on things on the surface. So perhaps the pressure from the "ocean of air" somehow produces the limit on how high water can be pumped. Parallel to what Stevin and Galileo had discovered about water pressure, this accumulated weight could support a column of water to a height that balances the weight of the ocean of air, but no higher!

Very often sophisticated developments (say, the Pythagorean theorem) entail logic far more complex than what I have just suggested for Torricelli. But sometimes there are clues directly available from Nature that put these major accomplishments within direct reach—not within the reach of anyone, of course, but within the reach of an Archimedes or Euclid. Looking ahead to the probability discussion, this was conspicuously true for the Pythagorean relation, since in any society advanced enough to build large structures (such as temples), the Pythagorean relation would be hard to miss. Someone is bound to notice that a simple way to make a right angle is to measure off a 3,4,5 triangle. And in any such society there is also (we can notice) always some development of mathematics. But then, sooner or later, someone will also notice that $3^2 + 4^2 = 5^2$. And surely, we may suppose, someone will be curious enough to wonder whether other such neat triples exist, and once these are looked for, 5,12,13 is too close by to be missed. Or someone will try some arbitrary right triangles to see how close they come to that striking relation, and will discover that as precisely as can be measured, *all* right triangles seem to exhibit the Pythagorean relation. Either approach requires nothing more than directly checking an obvious thing to check.[5]

In contrast, it is hard to see how Torricelli could find evidence that could be brought to bear on his "ocean of air" conjecture (or later in this chapter, how Pascal and Fermat could have managed the breakthrough to probability) without persisting in pushing beyond any direct signals from Nature. Consequently, a pragmatic way to define AC inquiry is as inquiry that requires a person to keep looking though there apparently is nothing in sight. For Torricelli, the naïve intuition would be that while air is a material thing (in extreme conditions it can blow your house down), it does not weigh anything. Even after Galileo managed to weigh it anyway, the intuition that it weighs *on* anything would not easily follow. Even if it were to occur to you that in some way it ought to weigh on you, it isn't obvious how you could detect that, since certainly you cannot feel it. Nor is it obvious why any of this would connect with what Galileo had learned (from his plumber) about the inability of pumps or siphons to carry water up higher than 34 feet.

Examples of discovery prior to 1600 not only are sparse, but they never seem to have this persistence-demanding character. Rather, they seem to have a falling-dominoes character. In appeals to experience from before c. 1600, it is typically easy to see what could prompt a clever fellow to take the step that brings the evidence into play, and that step having been taken, we can see what could prompt the next step.[6] The sequence (if there is a sequence) doesn't seem to require that stubbornness in the face of discouraging intuitions that is the mark of AC inquiry, where even a faint glimmer can sometimes be alluring enough to tempt a person to devote effort to exploring what could easily turn out to be a blind alley.

It takes an AC move to reframe the question of why pumps could *lift* water no more than 34 feet into a conjecture that the invisible weight of an ocean of air could only *push water up* 34 feet into a tube (imagined as empty of air). Torricelli had to think obliquely about the conjectured ocean of air in terms of a parallel to what Galileo and Stevin had learned about bodies of water. Air might be thought of as a fluid, but in contrast to water, it would be a fluid that can be compressed. Galileo and Stevin had shown that a point within a fluid feels pressure (weight per square inch) determined solely by the vertical depth at that point. But with some AC Case iii effort, a person could see how this surprising effect for an incompressible fluid with a well-defined column height might be applied to a conjectured compressible fluid of unknown column height. This is hardly an obvious thing to think of. And to directly test the idea would have required Torricelli to somehow start from an empty tube—empty even of air, which he could never achieve in practice even if the Aristotelians were wrong in claiming that it was impossible in principle.

So to review: If I hold a brick, I feel pressure on my hand from the weight of the brick. If I put a second brick on top of that, I feel twice the pressure. That is how common scales work. If air has weight, however slight, there should be parallel pressure at the surface of the Earth from the accumulated weight of all the air above, just as I feel the pressure on my hand from the bricks. What prevents pumping a column of water beyond 34 feet would then be just that 34 feet is as much of a column of water as it takes to balance the accumulated column of air.

But experiments with columns of water at least 34 feet high would not be very practical. Torricelli, however, did not fail to notice that if water can be supported to 34 feet, then with some fluid that is denser than water, experiments could be done on a smaller, more manageable, scale. Mercury is so much denser than water that a column of mercury 30 inches high should behave like a column of water 34 feet high.[7] Nor (another AC move) did Torricelli fail to notice that he did not need to create a column with nothing in it (a vacuum) to see how far his hypothesized weight of the ocean of air would push a fluid up. He could do the reverse: Start with a *filled* tube of fluid, upend it in a basin, and see how far the fluid fell.

Once all that has been done, it seems obvious. Until it is done, only someone who had thought hard about where he might look next would be likely to think of it. But now a manageable and direct test could be arranged, and Torricelli got the result that he expected and more. For simple variations of this inverted-tube setup made it hard to offer a coherent alternative to his account.

If the parallel with the Stevin and Galileo results with water holds, then if the tube is tilted, this should not change the vertical height of the column that would balance the pressure of air outside the tube against the pressure of mercury inside. Therefore, mercury should be pushed up into the tilted tube to maintain the vertical height of the column, reducing or eliminating the empty space at the top of the tube. Or if the level of the mercury in the basin was raised by adding more mercury, the same thing should happen: Mercury should move from the basin uphill into the tube. Experiments could not be much simpler: Turn the tube into the basin, release the plug, and note how far the mercury falls. Tilt the tube to the side, or set it deeper into the basin. The effects are all very striking. Slightly more elaborate arrangements yield even more striking effects. But it does take alertness to AC moves to put an experimenter into a position to do what can so simply be done. Here Torricelli could expect that the height in the tube would be determined by the weight (per square inch) of the atmosphere. He could weigh the atmosphere with a tube of mercury!

This yielded the then astonishing claim that we are subject at every moment to enormous pressure. But we are entirely unconscious of it. If you hold your hand out, exposing the 12 square inches of its surface, you are holding up a weight of about 12 × 15 = 180 pounds. But you feel nothing! Why not? Because the underside of your hand is being pushed up by the same 180 pounds of pressure. But then why is your hand not crushed by this 180 pounds of pressure converging on it from both sides? Because everything in your body is under the same pressure, so the pressure inside your body is pushing out as hard as the surrounding air is pushing in. No net effect is perceptible. The forces that Torricelli claims are operating are huge, but at the same time imperceptible—like the blood that is racing around your body as you read this, or as the Earth under your feet is racing around the heavens.

Belief in Torricelli's claim was not easy. If something seems preposterous but unanswerable, as we have seen in other contexts, there will be a tendency to just ignore it and hope that someone else will find a sensible explanation that will make the nonsense go away. Therefore, yet more experiments turned out to be needed to fully settle the issue.

This finally brings us to the celebrated mathematician-scientist-philosopher Blaise Pascal, whose most famous experiment enlisted his brother-in-law to pack a container of mercury, a basin, and a glass tube to repeat Torricelli's test at the top of a mountain.

Pascal reasoned that since air is compressible, at the bottom of the sea of air it will be squeezed by the weight of all the air above it. Hence air at the Earth's surface must be far denser than air far above in the sky. Therefore, if we could move up into the atmosphere even a modest amount (relative to the whole extent of the atmosphere, whatever that might be), we might rise above a nontrivial fraction of the whole amount of air in the atmosphere. Then the Torricelli experiment that supported this ocean-of-air claim would show a noticeably lower weight of air atop a mountain. That is how the Torricelli experiment came to be repeated on top of a mountain. And, as Pascal hoped, the mercury fell several additional inches.

Many other experiments were done that in the aggregate put Torricelli's claim beyond reasonable dispute. Of course, it is never possible to end all doubt. Hobbes never accepted the "ocean of air," leading to a famous controversy with Robert Boyle.[8] But here our concern is only with the obliqueness of Pascal sending his brother-in-law up the mountain. Evident as Pascal's experiment might look to us today, it was a decidedly AC move then. It provides an example of AC inquiry Case iii: A chain of argument leads to an inference that if the "ocean of air" account is correct, then the mercury should fall farther on top of a mountain. Having seen that far, it becomes directly apparent what to do.

Perhaps this sequence seems easy enough, looking back, to be hardly distinguishable from direct inquiry. But at the time supporters of Pascal and of Descartes got into a bit of a priority dispute over who had thought of the experiment first. The merits of the rival claims are irrelevant here, but that there could be such a dispute—and between the two most brilliant investigators in France—is enough to show that however obvious this experiment might look to us, it was not obvious then. No one but Pascal and Descartes laid claim to this idea.[9]

And a few years later, before turning his mind entirely to religion, this same Pascal, now an experienced user of AC inquiry, became involved with Pierre Fermat (of the famous "last theorem"[10]) in correspondence about puzzles that a mutual friend inclined to both dicing and mathematics was moved to pose but could not solve.

Step by Step to Probability

Fair odds (or their equivalent) were never in question for simple gambles like tossing a coin (1:1) or rolling a particular point with a six-sided die (5:1). But suppose we know fair odds of winning each of a pair of gambles. Could we then calculate fair odds of winning both gambles? Could we calculate fair odds for winning at least one of the pair of gambles? The emergence of mathematical probability from Pascal, Fermat, and Huygens—the three best mathematicians active at the time, Newton still being a bit too young to qualify—consisted just of seeing how to handle the simplest cases of this sort, where fair odds for the individual gambles were not in question, and where what came to be labeled independence between the individual gambles was taken for granted.

Suppose a match is interrupted short of the required winning number of points. Assume that each player has an equally good chance to win any point. How should the stakes be divided such that neither player gains an advantage from the interruption? The fair price would be the fraction of the prize that a competent player would willingly pay to take over a position, or the fraction that would seem fair to such a player to sell a position. The fair fractions to leading and trailing players should of course add up to 1 (no money should be left on the table), and fair odds would be just the ratio between the two fractions. This was the famous "problem of points," which had been discussed intermittently for some three centuries, without a resolution.

Suppose Smith needs 4 more points and Jones needs 7. Then a bit of counting will show that a winner will be determined after a minimum of 4 more points (if Smith wins 4 in a row), and within a maximum of 10 more points (if Smith wins 3 and Jones 6 of the first 9 throws, after which the tenth point will decide the winner). Using $\{x, y\}$ to mean Smith that needs x points and Jones needs y points, we can label this case $\{4, 7\}$. The sequence that would conclude the match could be as short as x points, or as long as $x + y - 1$ points. And we want to calculate a fair division of the prize.

But if the score is tied ($x = y$), certainly the only fair division will be equal. For each player in this simplest case, the fair price for buying or selling a position must be one-half the stake. So without any calculation, we know the division for the simplest case, $\{1,1\}$. We can move on to the next simplest case, $\{1,2\}$, where either Smith will win the next point (hence immediately win the prize), or the game will be tied. But we already know the fair result for both these possibilities. For the first, since Smith has won, of course she gets the whole stake. In the alternative there is a tie, so Smith can claim half the stake.

Fair division then comes out as 3/4 for Smith, since Smith can equally easily either win the whole stake (if she wins the next point) or win half the stake (if she loses the next point). This same reasoning covers any case of the form $\{1, y\}$. For $y = 3$, for example, at the next point Smith has equal chances of winning outright or of reaching $\{1, 2\}$. Again, we already have a fair division calculated for both cases. The fair share for Smith would be 7/8: half of an outright win plus half of the 3/4 value of $\{1, 2\}$. And in general, the correct share to Smith for any case $\{1, y\}$ would be $[(2^y - 1)/2^y]$.

You may not have quite noticed, but we already have here the emergence of probability, which had been lying dormant for 2000 years. No new ideas are needed to reach any other situation, step by step, by continuing this simple procedure. For the next simplest case, $\{2, 3\}$, one more play will make the match a tie (if Jones gets the point) or makes it $\{1, 3\}$ (if Smith gets the point). But as before, we already know values for those cases. By the reasoning already in hand, the correct share for Smith is half of each: half of 1/2 of the whole stake, plus half of 7/8 of the stake. So the fraction of the prize to Smith is 1/4 + 7/16 = 11/16, leaving 5/16 for Jones.

Probability and AC Inquiry

But if the breakthrough on fair division was this simple, that makes the puzzle of how the discovery that odds for compounded gambles can be calculated should have been so long delayed. The situation seems so odd that it is probably a relief to notice that there is a difficulty in this simple analysis. The difficulty is that an alternative simple analysis leads to a different result. And the alternative analysis has an advantage: like various Aristotelian physical intuitions discussed earlier, it appeals more directly to familiar experience. The alternative was first pointed out by Gilles Roberval, holder of the chair of mathematics at the University of Paris. Even decades later, a more famous mathematician (D'Alembert) was still using a version of Roberval's argument. And a bit later in this chapter, you will have an opportunity to discover a variant for yourself.

Consider the {1, 2} case again. If you look to what a person could actually encounter in the world, there are just three possible continuations of the match. Let S mean a point for Smith and J a point for Jones. If we play the match out, the three actual possibilities are that Smith wins the next point (and with it the match), or Jones wins the next point but Smith wins the third and deciding point, or Jones wins the next point and also the deciding point. Of the three possibilities (S, JS, JJ), Smith wins 2 of 3. Why, then, isn't Smith's share 2/3 of the stake, not the 3/4 that Pascal and Fermat give her? How could Fermat and Pascal be sure that they were right? How could they persuade others that they were right when if you just directly count what you can see in the world, it is Roberval who is right?[11] And of course in defending their argument for 3/4 as the correct share to the leading player, Pascal and Fermat obviously could not invoke later knowledge of the calculus of probability that was in the process of being invented right there.

We are in the sort of situation described in Chapter 3, where it is not apparent how to firmly establish what makes more sense. For the "problem of points," until a convincing solution had been shown, how could anyone be certain even that a definite solution to this puzzle existed? And that is where things were stuck for a very long time.

An Around-the-Corner Move

There was an always-available route to the emergence of probability, just as there was (since Ptolemy) an always available route to the Copernican discovery. But for probability as much as for the crucial Copernican insight, the always available route was apparently very difficult to see—otherwise, why did it take so very many centuries before anyone noticed?

Recalling the earliest AC move I had occasion to consider, Stevin (Figure 5-1, from the mid-1580s) wanted to show how the mechanical advantage of an inclined plane relates to the angle of inclination. He noticed (around the corner) that the situation could be made transparent by considering weights that did not have any effect at all on the balance of forces. And parallel to that, going around another corner, the decisive move here for Pascal and Fermat was noticing the usefulness of counting cases that would never actually occur.

But for someone who could find his way around the corner, things could go very nicely. The Fermat-Pascal correspondence reflects a certain element of high spirits in seeing how prettily things are working out. For a person—and certainly a person as perspicacious as either Pascal or Fermat—could scarcely start down this road without noticing that many things come into view beyond the narrow issue of a solution to the old problem of points.

Think again about the simple {1, 2} case, where Smith needs just one point and Jones needs two. Suppose Smith wins a point if the roll of a die turns up an odd number, and Jones wins if the number is even. At most, two rolls will be needed to finish the game. We could use as many dice as are needed to represent each possible roll—here just two: say red to represent the first roll and blue to represent the second. The red and blue dice can be rolled together, saving a bit of time even for this simple case and a lot of time for more complicated cases. Using capitals for an odd number (point for Smith) and lower case for an even number, the possible rolls are RB, Rb, rB, rb. Smith wins 3 times out of 4, losing only in the last case. And you can see that neither Smith nor Jones gains any advantage by playing out the maximum number of trials that might be needed, even if the match is already decided. Smith wins only when she would have won if the match stopped when it was decided (as in actual play it would).[12]

With all the possibilities in front of you at once, you would be physically confronted with the error in Roberval's 2/3 count, where only sequences that might actually be played out are counted: here, only R, rB, rb. But the single case in which Jones wins (rb) takes two favorable rolls in a row for Jones, whereas the R case (in Roberval's three-item list) takes only one favorable roll for Smith, and is twice as likely.[13]

Taking notice of that brings us back to the Fermat and Pascal solution, where the leading player's share is 3/4. This is simple and yet quite astonishingly tricky. You will have an opportunity to test your own intuition in the next section.

Extending this "count all cases—even if they will never actually occur" procedure to the {2 ,3} case, the maximum number of rolls to determine a winner is now four. You can see how to count the ways for the four dice to fall without having to actually roll them. Jones would win the match only if the dice produce at least three out of four even numbers. Suppose the dice are now colored red, green, blue, and yellow. Again with lowercase letters for even numbers and capitals for odd, we can just count all (five, it turns out) winning rolls for Jones: rgby, rgbY, rgBy, rGby, Rgby. In all other cases Smith wins. And it is not hard to see (or, for a small number cases, just to count) that a total of 2^n possible sequences are possible: here $2^4 = 16$ possibilities.

So for the case at hand, the winning fraction for Jones is 5/16 (since Jones wins just five of the sixteen possible rolls of the four dice), leaving the balance (11/16) for Smith, just as the original Pascal/Fermat argument claimed. Now, however, a firm answer to Roberval's rival claim is in hand. And you can also now see what apparently makes the problem tricky: To get it right, you have to use the AC move of counting cases that would never occur!

And you can now notice that it is unnecessary to count the possibilities one by one. All that is needed is the fraction of throws with four dice that yield at least two even numbers, and the examples already in hand are enough to suggest a quick formula for "counting without counting."[14] In fact, the quick formulas had already been worked out (repeatedly) in earlier discussions of combinations that had nothing to do with probability. So we move here, as Pascal and Fermat moved four centuries ago, very quickly from step-by-step counting to a much more efficient procedure that in turn opens the door to dealing with much more complicated questions. Fermat and Pascal saw that accumulating further results this way would be unnecessarily tedious. With these simple cases in hand, modestly sophisticated mathematics would carry a person well beyond where we have gotten, to deal with cases where independence does not hold, where the chances at the next toss are not equally likely, where there are more than two possibilities at each point, and so on. Although it took 2000 years to reach the simplest calculations for compound gambles, within a few decades an elaborate calculus of probability had developed.

A Cognitive Illusion

We are now logically past the difficulty posed by the argument that in the {1, 2} case, 2/3 and not 3/4 is the fair division for the leading player. But a significant element of this story comes from noticing the force that illusory probability intuitions still have today, and even among people with far more than average experience with probability calculations. That versions of the illusion are still hard to escape should warn us against assuming that what Pascal and Fermat did should have been easy. Perhaps you would like to try this problem yourself:[15]

There are three poker chips in a cup. One is marked with a blue dot on each side, and another has a red dot on each side. The third chip has a blue dot on one side and a red dot on the other. So there is one blue/blue chip, one red/red chip, and one blue/red chip.

Without looking, you take out one chip, and lay it on the table.

1. Suppose the side that is up turns out to be blue? What is the chance that the side that is down will also be blue?
2. What if the side that is up is red? What is the chance that the side that is down will also be red?
3. Before you see how the chip has fallen, what is the chance that it has the same color dot on both sides?
4. Suppose you answered 1/2 in response to questions 1 and 2. That would mean that whichever color of the chip is up, the chance is 50/50 that the color on the down side is the same. But if at question 3 you said that the chance is 2/3, aren't you contradicting yourself?

The usual response to questions 1 and 2 is indeed 1/2. As implied by question 4, the usual response to question 3 is 2/3. And the response to question 4 from the large majority who report 1/2 for questions 1 and 2 but 2/3 for question 3 is most often that there must be some mistake in the reasoning that claims to show a contradiction. Indeed, this is very often a most emphatic response, which is hard to overcome for people whose experience in the world gives them confidence that they could not have a mistaken clear intuition about such a simple matter. Nevertheless, you will eventually come to agree, however strong your intuition otherwise is right now, that the correct answer is always 2/3. The powerful intuition that 1/2 is right for questions 1 and 2 is an illusion.

If you confidently see 1/2 as the correct response to questions 1 and 2, you are only doing what distinguished predecessors like Roberval and D'Alembert did centuries ago (along with countless undistinguished predecessors): You are failing to distinguish cases that are physically distinct but somehow hidden in the way the problem presents them to you. To solve the problem of points correctly, you have to see why it is necessary to count cases that in practice will never occur. Here, if the upside of the chip that you pull out of the cup is red, then you notice (of course) that that chip must be either red/red or red/blue. And it is then remarkably difficult to escape the intuition that the probability of each is 1/2. But suppose the two red dots on either side of the red/red chip have been secretly labeled *a* and *b*. Now there are certainly three possibilities when you see a red dot on the upside, not two: red/blue, red^a/red^b, and red^b/red^a. All are equally likely, even though for the red/red chip, two physically distinct cases are perceptually indistinguishable. There is a double chance that the chip you have pulled is the red/red not the red/blue.[16]

Many readers will still be left in doubt at this point, which provides a further use for the puzzle. For it is not so easy even today to prompt someone to look for around-the-corner evidence. The example at hand turns on the conflict between a strong intuition that happens to be wrong (that the "chips" probability is 1/2) and an argument that purports to show that the correct answer is something else (in this case, 2/3). But since everyone today is familiar with the usefulness of running experiments, isn't it obvious that the question can be settled very quickly by just running some trials? Several dozen trials could be run in just a few minutes. So if the correct answer is indeed 1/2, then it ought to be easy to support that by an experiment directly checking whether the fraction of red/red when the up color is red is closer to 1/2 or closer to 2/3. But the only people who ever propose running such an experiment are those who have already escaped the illusion and are trying (commonly with difficulty) to persuade someone else. It simply never occurs to people who have not yet seen the error in the usual intuition that in a few minutes' time a convincing answer could be obtained by just watching what happens in a series of trials. It has to occur to a person to take attention away from the single case probability of the question as posed and consider the parallel question for a long series of such questions. That is so even though since ancient times people seem to have understood that the two questions must have a common answer. If this is actually tried, it turns out to be unexpectedly easy to dispel the illusion, since a person once is in the physical situation usually quickly sees that the probability must be 2/3.

But this quite simple AC move is routinely missed today. Noticing that, we ought to be less surprised that this kind of move (what I have been calling an AC move) might not have been easily available—that it was indeed around-the-corner—four centuries ago.

Another Side of Probability

But an experiment or gamble that that turns on tossing symmetrical coins or dice is highly atypical of our total experience with probability. We cannot go through a day without making choices that depend on assessment of risk, and they rarely look at all like tossing coins or rolling dice. Which line at a supermarket checkout shall we wait in? Do we need to bother refilling the gas tank yet? Will the IRS audit if we take a deduction this large? And a thousand other such situations. Usually we have no more than a comparative sense of what the chances may be. Consequently, before the emergence of probability theory, the intuitions about probability that would have been guiding someone would have made it very hard to think of chance as a countable fraction of favorable cases over a long run of observations. That is very far from the usual way we encounter probabilities in everyday life.

And though counting is hardly an invention of the Scientific Revolution, what can be called "inverse counting"—looking, necessarily around-the-corner, for some way to attach numbers to something that was not previously seen as quantitative—does seem to be an invention of the Scientific Revolution. The simplest explanation of why the emergence of probability was delayed an extra half century (relative to other long overdue discoveries that appeared in the burst around 1600) is that here a further difficulty was the novelty of this particular sort of move.

Over the decades since 1600, several striking examples had emerged of clever ways to attach numbers to what had never been treated that way, emerging not in a deliberate move of that sort but pretty well forced by the ongoing work. It is not clear that Galileo quite got to instantaneous velocity, but at a certain point it becomes clumsy to avoid.[17] Atmospheric pressure could hardly avoid coming to be measured by inches of mercury. Probability (as a number) itself has something of this kind of forced emergence. Pascal, Fermat, and Huygens all held onto what we would call expectations (prize × probability) where there were familiar units (of money) to count. Treating probability alone as a meaningful thing only gradually took hold.[18]

Even in actual gambling, until tools for that became available with the breakthrough work discussed in this chapter, there was no salient need for a way to calculate fair odds. Outside the prescribed (and relatively modern) context of a gambling casino, players ordinarily take turns on either side of a gamble, and therefore it doesn't matter whether the gamble is fair. There is no urgent reason to calculate correct odds, and hence no salient motivation to work at figuring out how to do that. Where a probability effect was so simple that it required no calculation of compound gambles, the puzzle was solved. Seeing, for example, why seven is the most common roll with two dice requires only directly counting the number of ways a pair of dice can fall. Many people did that sort of thing long before Pascal and Fermat. But a quite trivial problem in compound odds, like the problem of points, had a solution that was around-the-corner. And on the record, until our Copernicans c. 1600 had shown the way, solutions around-the-corner remained out of reach.

But once probability was seen as a pure number for cases of dicing, where (after Pascal and Fermat did it) it is easy to see what to count, discussion moved remarkably quickly to probability where quantity was nowhere in sight. Within a very short time, Pascal was discussing his famous wager on the probability that God exists, and less religiously committed figures were beginning to think of the probability that a man would die next year, or that a ship would safely complete a voyage.[19]

This suggests that the emergence of probability might instructively be seen as being itself an AC move on a far grander scale than has so far been considered. Neither Fermat nor Pascal seems to have been much of a gambler. So why were they deeply enough engaged by questions from their gambling friend (Chevalier de Mere) to work their way to a discovery that had been missed for 2000 years? The rapidity with which their resolution of trivial questions about dicing was extended to the vastly larger stage of what grew into the modern science of statistics suggests that perhaps they at least tacitly recognized what today is taken for granted: that the best route to coming to grips with actually important questions about probability would start (around-the-corner) by dealing with something that was trivial but quantitatively unambiguous: well-defined questions about fair odds and gambling.

Notes

1. A person trying to lift the bucket fast encounters resistance from the viscosity of the water that has to be pushed out of the way.
2. Galileo's density of air relative to water (1:440) was off by almost a factor of 2, but it was a vast step forward from a situation in which it was disputed whether air had any weight at all.
3. Jerome Cardano is commonly reported to have anticipated this result, weighing the air about 1560. But if he did make this important discovery, he seemed to have forgotten it when he wrote his autobiography, nor was it published anywhere until an edition of his works marred by many editorial errors was published in the 1660s. A parallel comment applies (Margolis 1993, Chapter 6) to a parallel claim for Cardano's priority as discoverer of a calculus of probability.
4. Galileo, like Gilbert and later Darwin among others, was energetic in finding out what artisans and other practical people knew from expe-

rience. It would have been entirely in character if he had been helped to his crucial insight into how to select lenses that could make good telescopes by careful questioning of lens makers in Venice and Florence.

5. Galileo (*Dialogue*, p. 51) remarks that "it is certain that Pythagoras, long before he discovered the proof . . . was sure that the square on the side opposite the right angle in a right triangle was equal to the squares on the other two sides. The certainty of a conclusion assists not a little in the discovery of its proof." And we now know that Galileo was right in his confidence that the Pythagorean theorem was known empirically before it was formally proven. Now-deciphered Babylonian and Egyptian inscriptions show that the result had in fact been known for something like a thousand years before the Greek invention of geometrical proofs.

6. For example, consider some important ancient discoveries. What is the direct evidence for supposing that the world is round? (1) Sailors knew that you could see land sooner from a lookout atop the mast than from the deck. (2) Travelers knew that as they traveled toward the equator, the pole star sank lower toward the northern horizon and stars that had been barely visible on the southern horizon now were seen higher in the sky. (3)When there is an eclipse of the Moon, the shadow of the Earth on the Moon is round. We have no formula for determining when evidence aggregates to enough of a case to settle matters. But it is certainly not surprising that this package of evidence proved to be enough.

   And on awareness that objects accelerate as they fall: Have you ever tried to catch a heavy object dropped from a balcony, in contrast to the same object dropped a much shorter distance? Or have you noticed that in (say) a waterfall, what had been a solid stream is stretched out into pieces as it falls—but how could that happen if the water was moving as fast at the bottom as when it first started to fall? And on the question of the loss of weight by an object under water, I have mentioned the common experience of lifting an anchor.

7. At first Torricelli could not get a glass tube (so that he could see how far the mercury fell) that did not burst under the weight of the mercury. Therefore, there were intermediate steps (with honey, for example) before the experiments with mercury, and also of course direct and practical experiments with making stronger glass tubes.

8. This is the focus of Shapin and Schaffer's well-known book (Shapin and Schaffer, 1985). I give a very different view of that controversy in Margolis, 1993, Chapter 11.

9. According to some accounts, it possibly was the third most famous French scientist of the period, Mersenne, who first thought of this, which hardly undermines the point.

10. Fermat famously jotted in the margin of a book that he could prove there were never triples (parallel to Pythagorean triples) such that $x^n + y^n = z^n$ for $n > 2$. Simon Singh (1997) provides an accessible account in his *Fermat's Enigma*.
11. You might get some further sense of Roberval's difficulty by thinking of whether a couple is more likely to have two boys as against a boy and a girl. In fact families with two children have a boy and girl pair twice as often as they have two boys (or two girls). The chance of boy, boy (or of girl, girl) is not 1/3, although the set of possibilities is just two boys, two girls, or one of each. This commonly strikes people as counterintuitive, as various physical effects do also, especially outside the context of a physics course where a person is drilled into seeing what physicists have learned since 1600. See the "chips" example later in this chapter.
12. Or a familiar example for Americans: if the World Series were always played out to seven games, it would have no effect on who would win the series: which of course is why the series stops as soon as one team wins four games.
13. And the resolution of the boy/boy versus boy/girl puzzle is the same. For parents to have two boys, they must have a boy as their first child, and then a boy again as their second, with probability of $\frac{1}{2} \times \frac{1}{2} = \frac{1}{4}$. But to have a boy/girl pair, the first child can be either sex, with a probability of 1/2 that the second will be opposite.
14. $N = n! / k!(n - k)!$
15. You might be familiar with the Monty Hall puzzle, which is a close kin to this "chips" problem. Essentially the same analysis applies to both.
16. This is taken from Chapter 3 of my *Dealing with Risk* (1996), which contains a detailed treatment of this trivial but often very stubborn puzzle, together with a discussion of its implications.
17. Drake was sure that he did not. See his introduction to Galileo's *Discourses*.
18. For a detailed account stressing this aspect—and overdoing it, I have to say, in the light of the evolution reported in this study—see Chapter 6 of my *Paradigms and Barriers*.
19. The history of insurance, annuities, and so on goes back to ancient times. Therefore, we are talking here only of the beginnings of actuarial calculations and the like, which by rationalizing these activities were able to make them far more efficient.

Epilogue

Cognition and Politics

Until the Copernican system had completely swept aside sophisticated opposition, the Tychonic compromise discussed in Chapter 2 provided a refuge for people who could not believe Copernicus but also no longer felt comfortable with Ptolemy. After the Church condemned Copernicanism, the Tychonic choice became the only option open to loyal Catholics. But how people came to see that option as a *sensible* compromise is odd.

As described in Chapter 2, a crucial element of Tycho's system was that for the model to fit with observations, the orbit of Mars must intersect the orbit of the Sun. That seemed to make that system unworkable in a world in which the heavenly bodies are carried by Ptolemy's solid spheres. However, Tycho said, he could prove that the spheres did not exist, opening the door to his superior version of geocentric astronomy.

Without exception, everyone since has accepted that, from Kepler through to the very best historians in our own time. Tycho's claims were treated as so obviously right that they needed no detailed argument. But it turns out that this confidence was triply misplaced. We have gone through that in Chapter 2. There actually was no compelling argument against the solid spheres available at the time Tycho made his claim. This, however, turns out to have been irrelevant to the Tychonic system, since Tycho was wrong to suppose that his system would not work in a world of solid spheres. And then the supposed superiority of Tycho's version of geocentric astronomy turns out to mostly reflect a lack of interest in considering alternatives, since a better alternative was almost trivially available.

How a writer settles on a project is often a bit mysterious, but here, I suppose, it is merely odd. By a kind of academic accident I realized that the conflict between Tycho's system and a world of solid spheres was in fact only an illusion, akin to the illusion that makes us all see the line with an outward V as longer than the line with an inward V in the Müller-Lyer drawing (Figure E-1). With the help of a simple move (cutting Tycho's own diagram into two pieces to make Figure 2-2),[1] the illusion becomes transparent. The move is analogous to putting a ruler against the two lines of Figure E-1. But that simple move was not simple to see.[2]

Once the illusion has been broken, Tycho's claim becomes much closer to being obviously wrong than to being obviously right. Indeed, I gradually realized that practically everything that was usually said about the Tychonic system was wrong. The evidence is very

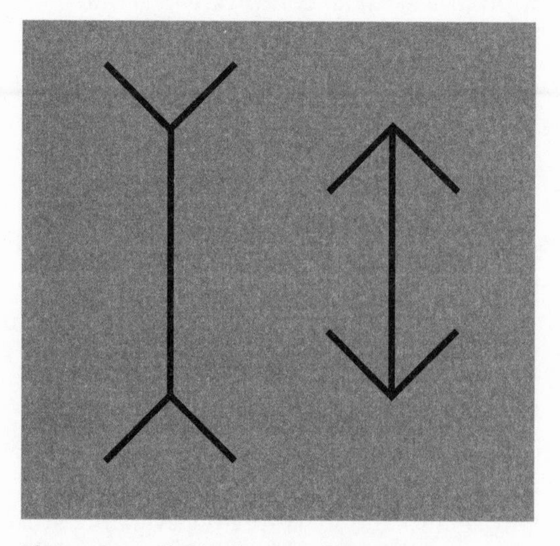

Figure E-1 The Müller-Lyer illusion

strong, as reviewed in Chapter 2. But noticing this decidedly strange situation prompted curiosity about what else was going on at the close of the sixteenth century that was connected to this strange behavior. And of course there was something else going on, and it involved the same sort of people who concerned themselves with rival systems of the world. I started to write a book under the title *Tycho's Illusion*, which was intended to use the Tychonic story as the point of departure for a discussion of cognitive effects in contemporary politics. As you have seen, I was diverted. But I don't want to close before at least introducing that contemporary theme.

I write not as a historian of science but as an academic interested in social choice, with an appointment in the University of Chicago's school of public policy. I serve on the editorial boards of journals appropriate to such an appointment, and I write books and articles in these fields.[3] One of the major puzzles of social choice is why people so often acquire passionate commitments in politics that seem unrelated to—or even in conflict with—any reasonable interpretation of where their objective interests lie. These puzzles, in turn, lead to more fundamental questions about how human judgment actually works, which is sometimes in contrast to how we think it ought to work.

But it is hard to study belief directly in the realm of politics. Passion gets in the way, and the choices are complicated. Consequently, I was led into areas where things can be made clearer (an "around-the-corner" move, I have to note), such as the cognitive experiments to which undergraduates taking a psychology course are subjected, but also episodes in the history of science at moments of what Thomas Kuhn (1971) memorably labeled *paradigm shifts.*

In science, far more readily than in politics, it eventually becomes clear where the more reasonable case lies. But along the way, persuasion, belief, and (here) discovery sometimes follow routes that are hard to fit into a view of science as a kind of codification of straight thinking. Episodes in science (especially early science, before things get too complicated) turn out to be particularly fruitful for understanding how progress is made or missed, and how belief takes hold and what it takes to change it. Unlike experiments with undergraduates, we are plainly dealing with exceedingly well-qualified actors who are conscious of their historical role and are making carefully studied choices about matters that are terrifically important to them. If we find anomalies of judgment here, we can be sure that we are dealing with features of human cognition that have real bite.

It has been that focus on psychology that led here to a new account of the Scientific Revolution. The argument has been concerned at every step with what could plausibly be going on in people's heads that can account for what they do and say. But now I want to return to the concerns that originally prompted this project.

That Tycho's illusion could have persisted for 400 years, repeated by all the most famous commenters from Kepler to Thomas Kuhn (and all the less famous as well) surely must tell us something about human cognition. We are now three decades into a prolonged debate among psychologists about rationality and human cognition. The debate has been mostly about how to interpret what appear to be stubborn cognitive illusions, analogous to the more familiar and undisputed perceptual illusions, as in Figure E-1.

Suppose we set aside cases of bad judgment that can reasonably be attributed to inexperience, fatigue, carelessness, ambiguity, bias, mere stupidity, or whatever else you might care to add by way of explanations of why human beings sometimes hold and defend illogical beliefs. After allowing for all that, does there remain solid evidence that under some conditions human intuition simply misbehaves? Could it actually be a property of how our minds work that smart people can sometimes produce stupid judgments, even when they are dealing with a matter that they know a good deal about, and when they are confronted with no apparent difficulty in bringing their competence to bear? Are there cognitive illusions parallel to the perceptual illusions that no one disputes? Under what conditions do they occur? Under what conditions can they be overcome? And are they significant enough to have important social consequences?

A complication here is that *rationality* is not, after all, just another technical term; it is more like *decency*, *beauty*, and *justice*. To say that someone is not on the side of rationality is not nice. Therefore, the label *rationality debate* comes mostly from the usage of people who see themselves as defenders of human rationality, not from those who think the evidence is that we are indeed vulnerable to cognitive illusions. I reveal which side I am on by a preference for talking about cognitive illusions, not about irrationality. I don't think human beings are irrational. I do think we are sometimes illogical, and more often so than we can notice.

Are there *socially significant* cognitive illusions, cases in which bad intuition has consequences that can play a significant role in how societies work? We are not talking here about illusions that are plainly rooted in passions, although that complication certainly cannot be avoided if we are talking about politics. The question here is not whether passions sometimes distort cool reason, but whether what seems to be cool reason itself is sometimes unreasonable, where passions (though surely never wholly absent) are quite clearly not a sufficient source of the difficulty.

I think the answer is that indeed our intuition is unreliable—meaning (of course) not that intuition is always wrong or even usually wrong, but that it is sometimes wrong even when it seems to us that it could not be wrong. And that can happen even when we eventually can come to agree that the logic that would readily have saved us from that was at hand, but that somehow it did not have the effect on our beliefs that we like to think that clear logic would have. We have a confident feeling of knowing we are right, but we are wrong. We have confidence we have carefully considered the case, but it turns out that something logically necessary was missed.

I wonder who could deny that this happens. Recall the "chips" discussion of Chapter 6. With mere puzzles, however, there remains endless room for argument about whether one or another of the special explanations (fatigue, carelessness, and so on) accounts for whatever error is at hand. We can agree that we are occasionally wrong on things we thought we could be sure about, but that does not settle the question of whether ordinarily competent judgment is sometimes systematically wrong on serious matters. The professional journals regularly provide fresh testimony that anything that some people (such as me) take as evidence of perverse intuitions can be interpreted (or rationalized) as just good judgment misled or misinterpreted.[4]

The cases that we would care most about, and that could provide the most striking evidence, are those in which the people making the judgment are (1) clearly competent individuals, (2) dealing with what they take to be an important question, and (3) with ample time to consider the matter. Of course, those are not the conditions of the tricky little questions put to undergraduates in psychology experiments. But the history we have been reviewing here provides a variety of examples in which such conditions hold. Tycho's illusion is the most striking, but we have also seen that many logically accessible discoveries waited 2000 years to be made, and that many bad arguments were casually accepted when they happened to coincide with what people are inclined to believe. These are cases in which the actors are as capable as human beings are likely to be of following a logical argument wherever it leads—and nevertheless the publicly expressed expert judgment of these actors is wrong! If Kepler and Kuhn were capable of missing easy logic, then perhaps you and I are too.

Any *current* controversy of this sort will occur in a rich social context that would make it difficult to cleanly isolate a claimed illusion. Even if I could persuade you that some belief you now hold is illusory, that would only *necessarily* persuade you that what you thought was a good reason for that belief was not in fact a good reason. That would not prevent you from then thinking of some other reason to replace the now-deflated one. In any real social context, there will ordinarily be some way to reformulate our reasoning that can rescue us from changing our minds. It is always possible to at least blur the difficulty to the point where we are no longer confronted with an unambiguous illusion. We are all pretty good at that.

We can do better with historical cases, and especially with cases from the history of science. Here we have many cases in which what had once been the focus of sharply conflicting judgments has now been settled to the satisfaction of anyone who understands the issue. No one any longer doubts that doctors can do terrible damage to women in labor if they do not wash their hands after leaving the autopsy room. Or that rocks sometimes fall from the sky (meteors). Or that it is possible to fly to the Moon. So if in some of these cases it is plain that a point was reached where logic lay on one side, but belief still went elsewhere, there is a good chance that we are looking at a case of illusory judgment, and a case where there might be a sufficient record to allow us to pin down just what might account for that.

What motivated this book was an expectation that a good deal might be learned by attending carefully to such cases. Though an extension of the argument to political and social choice is too ambitious (and too controversial) to adequately tackle in these concluding remarks, a reader who senses the importance of these issues by now has a substantial range of opportunities to learn more. My own previous book (Margolis, 1996) was on expert/lay conflicts of intuition on matters of environmental risk. John Graham and colleagues at Harvard (1995) have written on closely related matters. Two of my colleagues at Chicago—Richard Thaler of the School of Business (1998) and Cass Sunstein of the Law School (2000)—have written on anomalies in economic choices and in the courts. Robert Shiller of Princeton (2000) has written on "irrational exuberance" in stock markets. Jonathan Baron of Penn and colleagues at Harvard (2001) have written of anomalies in public policy. And more is available and in train.

These ideas have not yet had much impact on how things are done. But their time will come, and perhaps it is not far off.

Notes

1. In April 1998 I published a short report in *Nature*, from which Figure 2-2 is reprinted.
2. In an account for the e-journal *Psycoloquy*, I describe how this emerged from a suggestion from Owen Gingerich. URL: http://www.harrisschool.uchicago.edu/wp/sup-08.html

3. One (*Public Choice*) deals with formal (mathematical) analysis of social choice, and another (*Journal of Policy Analysis and Management*) deals with public policy analysis.
4. The leading advocates of the two positions (as of 2001) are Daniel Kahneman (for taking illusions to be important) and Gerd Gigerenzer (for seeing emphasis on illusions as essentially misleading). They provide a good view of the debate, one that is still quite up to date as of 2002, in *Psychology Review* 1996. My own sympathies obviously are with Kahneman. Gigerenzer's work is extremely good. But it is a puzzle to me why he sees such conflict between two positions that are more often complementary than conflicting. But I have not succeeded in persuading him of that.

APPENDIX

AC and the Industrial Revolution

I started this book by noting that the notion of a Copernican revolution, or a Scientific Revolution (so of course of a connection between the two) is quite out of fashion in recent historical writing. There have been parallel doubts about whether there really was something properly termed "the Industrial Revolution," understood to mean some particular turn which proves to be profoundly important.

For the skeptics—entirely parallel to the situation for the Scientific Revolution—the difficulty in identifying anything special invites a simple resolution. Nothing revolutionary has been identified because nothing revolutionary happened. "The causes [of the Industrial Revolution] have been so difficult to agree on because there was no 'Industrial Revolution'; historians have been chasing a shadow." And "the absurdity is . . . that the term [Industrial Revolution] is taken seriously."[1]

For the Scientific Revolution, there is a common notion that it turned on an appeal to experiment. Parallel to that, for the Industrial Revolution, would be the notion that the Industrial Revolution turned on the application of science to practical problems. But for the Scientific Revolution, historians noticed that experiment did not begin with the emergence of modern science. And, for the Industrial Revolution, while eventually the application of science-based technology became essential to the sustained rise in productivity that began about 1760, historians noticed that the onset of the "revolution" preceded any important practical use of the new scientific knowledge. But if A followed B, A could not be the cause of B.

The great example is James Watt's radically improved steam engine. The consensus among economists is that the Industrial Revolution (marked by the emergence of a shift of work from cottage to factory) was underway several decades before Watt produced his engine. Rather than initiating the Industrial Revolution, Watt's own firm (Bouton & Watt) used the already emerging factory system to produce his steam engines. Yet on the stark evidence of Table I-1, there was indeed an identifiable and striking turn that marks the onset of what indeed proved to be a Scientific Revolution. I now want to suggest how the same AC/DC account that identifies something specific linking the Scientific Revolution to Copernicus might also connect the Scientific Revolution and the Industrial Revolution of the following century.

"A Taste for Science"

Early in the Industrial Revolution, a key figure remarked on the pervasive "taste for science, over all classes of men."[2] A few years before the conventional starting date of the Industrial Revolution (1760), Boswell reports that Samuel Johnson, "was some time with [his rather dissolute friend] Beauclerk at his house at Windsor, where he was entertained with experiments in natural philosophy."[3] But interest in the new science, even among people not themselves engaged in anything remotely scientific, dates back much earlier, to the very beginnings of what became modern science. Galileo's debate with his critics over floating bodies (see Chapter 4) was conducted as a kind of court entertainment. And in England, the Lord Chancellor (Francis Bacon) famously took an intense interest in science, and wrote at length to tell the scientists how it should be done.[4] By mid-century, the Royal Society was taking form as a kind of gentlemen's club of people interested in science.

This wide interest in the new science put something revolutionary in play long before anything visible enough to be called the Industrial Revolution could be noticed. For if you were intrigued by what people like Galileo and Gilbert and their successors were doing, you were likely to be influenced by the around-the-corner aspect of how they were doing it. But in science an isolated discovery can be a big discovery (Jupiter has moons, and so on); while in practical affairs, a conjunction of many things is ordinarily required to make something decidedly new decidedly important. And to produce results visible as a great change in an entire economy requires a conjunction of those conjunctions, and considerable time as well. The things that would be changed by the Industrial Revolution—organization of work, traditions and tools of work, habits of consumption, and so on—are not things that could be abruptly changed by one person as Galileo could startle Europe with his little book on the telescope. Specific major innovations, such as Watt's steam condenser and Bessemer's steel-making process, came only later on, after the Industrial Revolution was already visibly in place.

Consequently, it makes sense that it would take much longer for AC inquiry to bring about the Industrial Revolution than it did to spark the Scientific Revolution. It was a half century after Copernicus died before his effect on discovery became apparent, so it should not be very surprising that it would take two or three times that long for the discovery of discovery to show aggregate effects on the mainstream of the English economy.

Until recourse to AC moves became a common thing, advances in practical matters were pretty much limited to what could come out of incremental improvements to what was already at hand. But with AC, advances can come from work not immediately tied to anything practical. A person might be alert to promising ideas, even though just how an idea could be put to practical work remained to be worked out. We might now see effort given to what is on the side, oblique to any immediately practical results. This is commonplace now. We call it "research." A person takes attention away from what is directly visible (for an artisan or entrepreneur, from the business immediately at hand) to ponder, offline so to speak, some intriguing possibility whose prospects are still around-the-corner.

As this novel capacity to do things in new ways came into place, there would be an irreversible increase in the rate of innovation among economic actors. It would be subtle compared to the far more immediate and visible effects of what I have been calling the discovery of discovery in science, but in the aggregate it could be powerful enough to change the world.

Whatever else was needed was somehow most at hand in Britain. Initially the Industrial Revolution took hold only there. So while a propensity to AC inquiry may have been a necessary condition, plainly it was not sufficient.[5] But until AC was on stage, periods of rapid economic growth were like earlier periods of scientific promise. They appeared from time to time, but ran their course, and left the world much as it had been. A sustained takeoff did not occur. Then in the years around 1600 a sustained takeoff began in science. The sustained takeoff in human productivity that followed does not plausibly seem only a coincidence.

Longitude and John Harrison's Sea Clock

Dava Sobel's *Longitude* has made the story of John Harrison's amazing clock familiar. It provides a good illustration of AC inquiry at work in practical matters decades before the visible onset of the Industrial Revolution around 1760. Beyond people with a special fondness for clocks, Harrison is not remembered for any particular invention. He just built a terrific clock that could solve the problem of finding longitude at sea.

So did John Harrison, in the 1720s still a village carpenter, do anything that looks like the sort of thing all our Copernicans started doing around 1600? The practical problem he was trying to solve is easy to describe. If you were flown blindfolded from New York to Los Angeles, you would know you were out west as soon as you found that the local time was three hours behind the time in New York. All you would need is a watch set to New York time. A ship at sea can tell its local time. So if a ship could carry a clock set to its home port, the skipper would know how far east or west of that port he had traveled. But an error of one minute means an error of about sixty miles in navigation, at a time when it took six weeks to cross the ocean, and when a good clock even on land would scarcely keep time to within one minute per day. But John Harrison somehow built a clock that (on his own much later recollection) could keep time to within a second a month. It is hard to believe the clock was that good; but it was certainly the most accurate clock by far anyone had built until that time. It was good enough to encourage Harrison to move to London to pursue his huge ambition, and (more convincing) good enough to win him at least cautious support from the Astronomer Royal and from the leading watchmaker in London.

Does this remarkable story fit the AC account? To start, the national project to which Harrison was responding itself takes AC inquiry for granted. In 1714 Parliament offered a huge reward for someone who would invent a way to keep track of longitude at sea. Nothing about how to do that was specified, and nothing in particular appears to have been looked for: just that somehow, someone might figure out how to do it.

And it was taken for granted that experiments would be needed. A Board of Longitude was appointed. Its commissioners would judge whether the prize had been won. But they were also authorized to finance experiments that showed promise. Harrison's work eventually got essential support from that provision.

Further, the precision clock that Harrison designed in the 1720s was itself an AC move. Of what practical use was a clock that could keep time over long periods to within a second a day? Where was the value in such long-sustained precision? Harrison plainly had in mind finding longitude at sea. But this clock was kept in step by a long pendulum. It was hopeless for keeping precise time on a ship at sea. It made sense only as a step on some around-the-corner path to what was really needed.

And how could such a clock be built, anyway? Until there was an even better clock available by which to judge it, how could you know whether it kept time to within a second a day? If you were Astronomer Royal, no AC move would be required for that. An astronomer would know exactly when a particular star would move past a given line of azimuth relative to when it did that on the preceding night. And he would have precision instruments to observe that. That a village carpenter could also know that says something about the availability of scientific knowledge. And that he could use that information says something about the propensity of this village carpenter to look for AC moves.

Having no observatory and no precision instruments, Harrison used the heavenly clock anyway. He arranged a secure place to set his head so that he looked just past the left edge of a windowpane to the right edge of his neighbor's chimney. As his critical star came into view an assistant would begin calling the seconds on his clock, so that he would know the exact second on the clock at the moment the star abruptly disappeared behind his neighbor's chimney. If you think about the AC moves in science we have had occasion to review, you will see some kinship here. Once he had a clock that could meet this stringent test of precision, Harrison could proceed with his bold ambition, since he now had (an AC move) a clock accurate enough to serve as a standard against which to test whatever he needed to test.

Notes

1. From Joel Mokyr's (1999) Introduction to a collection of essays, though he cites these to illustrate a view he himself certainly does not share.
2. From James Keir's *Dictionary of Chemistry*, quoted in Musson and Robinson, *Economic History Review* 13:222–244.
3. Boswell's *Life of Johnson*, p. 176.
4. Bacon's ideas here are different from those of our Copernicans. He disapproved of his Court colleague Gilbert's style of science and thought Copernicus was wrong too. But he wrote eloquently of how it should be done. Keynes (1966) offers a crisp appraisal of this odd situation.
5. Douglass North (1981, Ch. 12) provides an account that stresses the development of better defined and more secure property rights as what favored the crucial acceleration of innovation in Britain. That may indeed have been what made Britain special. But if the conjecture here is right, that would be fertilizer on seeds planted throughout Europe.

References

Apian, Peter. *Astronomicum Caesareum*. 1540; facsimile, Leipzig, 1967.

Applebaum, W. (ed.). *Encyclopedia of the Scientific Revolution*. Garland, 2000.

Baron, J. *Thinking and Deciding*. Cambridge University Press, 2001.

Bruner, J. S. *Beyond the Information Given*. Norton, 1973.

Burnett, D. G. "The Cosmographic Experiments of Robert Fludd." *Ambix*, 46 (1999):113–130.

Butterfield, H. *The Origins of Modern Science*. Free Press, 1957.

Butts, R. E., and J. C. Pitt. *New Perspectives on Galileo*. Reidel, 1978.

Cohen, M. R., and I. E. Drabkin. *A Source Book in Greek Science*. Harvard University Press, 1958.

Copernicus, N. *On the Revolutions of the Heavenly Spheres*. 1543; A.M. Duncan trans., David and Charles, 1978.

Crick, F. *What Mad Pursuit: A Personal View of Scientific Discovery*. Basic Books, 1988.

Dobrzyki, J. "Nicolaus Copernicus: His Life and Work." In B. Bienkowska (ed.), *The Scientific World of Copernicus*. Reidel, 1973.

Drake, S. *Galileo at Work*. University of Chicago Press, 1978.

Drake, S. *Essays on Galileo and the History and Philosophy of Science*. 3 vols.; University of Toronto Press, 1999.

Drake, S., and C. D. O'Malley. *The Controversy on the Comets of 1618*. University of Pennsylvania Press, 1960.

Duhem, P, *The Origin of Statics*. 1905; reprint Kluwer, 1991.

Encyclopedia of the Scientific Revolution (*see* Applebaum).

Galilei, G. *Discourse on Bodies in Water*. 1612; translated by T. Salusbury, University of Illinois Press, 1960.

Galilei, G. *Dialogue Concerning the Two Chief World Systems*. 1632; translated by S. Drake, University of California Press, 1967.

Galilei, G. *Two New Sciences*. 1638; translated by S. Drake, University of Wisconsin Press, 1974.

Gilbert, W. *De Magnete*. Translated by S. P. Thompson, Basic Books, 1958.

Gingerich, O. "A Fresh Look at Copernican Astronomy." In *The Great Ideas Today*, Encyclopaedia Britannica, 1963.

Gingerich, O., and J. R. Voelkel. "Tycho Brahe's Copernican Campaign." *J. Hist. Astron.*, 29 (1998):1–34.

Graham, J. D., et al. *Risk vs. Risk*. Harvard University Press, 1995.

Hall, M. B. *Nature and Nature's Laws*. Harper Paperbacks, 1970.

Hellman, Doris. *The Comet of 1577*. Columbia University Dissertation Series, no. 510.

Hoffman, P. *The Man Who Loved Only Numbers*. Hyperion, 1998.

Hoffrage, U., et al. "Communicating Statistical Information." *Science*, 290 (2000):2261–2262.

Jardine, N. *The Birth of History and Philosophy of Science*. Cambridge University Press, 1984.

Kelly, S. *The De Mundo of William Gilbert*. Menno Herzberger, 1965.

Kepler, J. *Optics*. 1604; translated by W. H. Donahue, Green Lion Press, 2000.

Kepler, J. *New Astronomy*. 1609; translated by W. H. Donahue, Cambridge University Press, 1992.

Kepler, J. *Gessamelte Worke*. W. Von Dyck and M. Casper (eds.), vol 13. C. H. Beck, 1945.

Keynes, G. "Bacon, Harvey and the Originators of the Royal Society. In *Selected Lectures of the Royal Society*, vol. 2, pp. 113–128. Academic Press, 1969.

Koestler, A. *The Sleepwalkers*. Macmillan, 1965.

Koyre, A. "Galileo and Plato." *J. Hist. Ideas*, 4 (1943): 400–428.

Kuhn, T. S. *The Structure of Scientific Revolutions*. 2d ed. University of Chicago Press, 1970.

Lewontin, R. *The Triple Helix*. Harvard University Press, 2000.

Lindberg, D. C. *Theories of Vision from Al-Kindi to Kepler*. University of Chicago Press, 1976.

Longomontanus (Christen Sørensen). *Astronomia danica*. Ioh. & Cornelivm Blaev, 1622.

McMullin, E. *Galileo: Man of Science*. Basic Books, 1967.

Magini, G. A. *Novæ coelestivm orbivm theoricae congruentes cum observationibus N. Copernici*. D. Zenarij, 1589.

Margolis, H. *Patterns, Thinking and Cognition*. University of Chicago Press, 1987.

Margolis, H. "Tycho's System and Galileo's *Dialogue*." *Stud. Hist. Philos. Sci.*, 22 (1991): 259–275.

Margolis, H. *Paradigms and Barriers*. University of Chicago Press, 1993.

Margolis, H. *Dealing with Risk*. University of Chicago Press, 1996.

Margolis, H. "Tycho's Illusion." *Nature*, 392 (1998): 857.

Mokyr, J. (ed.) *The British Industrial Revolution*. 2d ed. Westview, 1999.

Musson, A. E., and E. Robinson. "Science and Industry in the Late Eighteenth Century." *Econ. Hist. Rev.*, 13 (1960): 222–243.

Neugebauer, O. *Astronomy and History*. Springer-Verlag, 1983.

North. D. *Structure and Change in Economic History*. Norton, 1981.

North, J. *Norton History of Astronomy and Cosmology*. Norton, 1995.

Pascal, B. *The Physical Treatises of Pascal*, Columbia University Press, 1937.

Pedersen, O. *Early Physics and Astronomy*. Cambridge University Press, 1993.

Popper, K. R. *The Logic of Scientific Discovery*. Hutchinson, 1959.

Porter, R. (ed.) *Reader's Guide to the History of Science*. Cambridge University Press, 2001.

Poulle, E. *Les instruments de la théorie des planètes selon Ptolémée : Équatoires et horlogerie planétaire du XIIIe au XVIe siècle*. Droz, 1980.

Randles, W. G. L. *De la terre plate au globe terrestre: Une mutation épistémologique rapide, 1480–1520*. A. Colin, 1980.

Randles, W. G. L. *The Unmaking of the Medieval Christian Cosmos, 1500–1760*. Ashgate, 1999.

Randles, W. G. L. *Geography, Cartography and Nautical Science in the Renaissance*. Ashgate, 2000.

Reader's Guide to the History of Science (see Porter).

Reber, A. S. *Implicit Learning and Tacit Knowledge*. Oxford University Press, 1993.

Rosen, E. "Copernicus and the Discovery of America." *Hispanic-American Histl Rev.* (May 1943): 357–370.

Sabra, A. I. *The Optics of Ibn al-Haythem*. Warburg Institute, 1989.

Schmitt, C. B. "Experience and Experiment." In *Reappraisals in Renaissance Thought*. Variorum Reprints, 1989.

Settle, T. B. "An Experiment in the History of Science." *Science*, 133 (1961): 19–23.

Shapere, D. *Galileo*. University of Chicago Press, 1974.

Shapin, S. *The Scientific Revolution*. University of Chicago Press, 1996.

Shapin, S., and S. Schaffer. *Leviathan and the Air-Pump*. Princeton University Press, 1985.

Shiller, R. P. *Irrational Exuberance*. Princeton University Press, 2000.

Singh, S. *Fermat's Enigma*. Walker, 1997.

Sobel, D. *Longitude*. Walker, 1995.

Sobel, D. *Galileo's Daughter*. Walker, 1999.

Stevin, S. *The Principal Works*. Swets & Zeitlinger, 1961.

Swerdlow, N. "The Comentariolus of Copernicus." *Proc. Amer. Philos. Soc.* 117 (1973): 428–512.

Swerdlow, N., and O. Neugebaruer. *Mathematical Astronomy in Copernicus's De revolutionibus*. Springer-Verlag, 1984.

Sunstein, C. (ed.) *Behavioral Law and Economics*. Cambridge University Press, 2000.

Thaler, R. *The Winner's Curse*. Princeton University Press, 1994.

Thompson, S. P. "Petrus Peregrinus de Maricourt and His Epistola de Magnete." *Proceedings of the Brtitish Academy*, 1906.

Thoren, V. *The Lord of Uraniborg*. Cambridge University Press, 1990.

Van Helden, A. *Measuring the Universe*. University of Chicago Press, 1985.

Westman, R. "Copernicans and the Churches." In D. C. Lindberg and R. L. Numbers (eds.), *God and Nature: Essays on the Encounter between Christianity and Science*. University of California Press, 1986.

Zilsel, E. "The Sociological Roots of Science." 1942; reprinted with comments by Wolfgang Krohn and Diederick Raven in *Soc. Stud. Sci.* 30 (2001): 925–949.

Index

I

J

K

L

M

N

About the Author

Howard Margolis is a professor in the Harris Graduate School of Public Policy Studies and the Fishbein Center for History of Science at the University of Chicago. He has held research appointments at the Institute for Advanced Study (Princeton), the Russell Sage Foundation, and the Massachusetts Institute of Technology, publishing extensively on cognition, public policy, history of science, and mathematical models of social choice.